T0320593

Principles of LED Light Communications

Towards Networked Li-Fi

Balancing theoretical analysis and practical advice, this book describes all the underlying principles required to build high-performance indoor optical wireless communication (OWC) systems based on visible and infrared light, alongside essential techniques for optimizing systems by maximizing throughput, reducing hardware complexity, and measuring performance effectively.

It provides a comprehensive analysis of information rate-, spectral-, and power-efficiencies for single- and multi-carrier transmission schemes, and novel analysis of non-linear signal distortion, enabling the use of off-the-shelf LED technology. Other topics covered include cellular network throughput and coverage, static resource partitioning and dynamic interference-aware scheduling, realistic light propagation modeling, OFDM, optical MIMO transmission, and non-linearity modeling.

Covering practical techniques for building indoor optical wireless cellular networks supporting multiple users, and guidelines for 5G cellular system studies, in addition to physical layer issues, this is an indispensable resource for academic researchers, professional engineers, and graduate students working in optical communications.

Svilen Dimitrov is a researcher at the German Aerospace Center (DLR) in Oberpfaffenhofen, Germany. He is involved as a project manager in the European project on Broadband Access via Integrated Terrestrial and Satellite Systems (BATS), aiming at the development of terabit/s satellite communication systems with optical feeder links.

Harald Haas is Chair of Mobile Communications at the University of Edinburgh, and Chief Scientific Officer of pureVLC Ltd. He first coined Li-Fi, listed in *Time Magazine's* 50 Best Inventions of 2011, and covered by international media channels such as the BBC, NPR, CNBC, the *New York Times, Wired UK, NewScientist*, and *The Economist*. His TED talk on the subject has been viewed more than one and a half million times, and in 2012 he received a prestigious Fellowship from the Engineering and Physical Sciences Research Council (EPSRC), UK. In 2014, he was selected by EPSRC and the Royal Academy of Engineering as one of ten RISE (Recognizing Inspirational Scientists and Engineers) leaders in the UK.

Principles of LED Light Communications

Towards Networked Li-Fi

SVILEN DIMITROV
German Aerospace Center (DLR), Oberpfaffenhofen

HARALD HAAS
University of Edinburgh

CAMBRIDGE
UNIVERSITY PRESS

University Printing House, Cambridge CB2 8BS, United Kingdom

One Liberty Plaza, 20th Floor, New York, NY 10006, USA

477 Williamstown Road, Port Melbourne, VIC 3207, Australia

314-321, 3rd Floor, Plot 3, Splendor Forum, Jasola District Centre, New Delhi - 110025, India

79 Anson Road, #06-04/06, Singapore 079906

Cambridge University Press is part of the University of Cambridge.

It furthers the University's mission by disseminating knowledge in the pursuit of
education, learning and research at the highest international levels of excellence.

www.cambridge.org
Information on this title: www.cambridge.org/9781107049420

First published 2015
Reprinted 2017

A catalogue record for this publication is available from the British Library

Library of Congress Cataloging in Publication data
Dimitrov, Svilen.
Principles of LED light communications: towards networked Li-Fi / Svilen Dimitrov,
Harald Haas.
    pages   cm
Includes bibliographical references and index.
ISBN 978-1-107-04942-0 (Hardback)
1. Optical communications.   2. Wireless LANs.   3. Light emitting diodes.
I. Haas, Harald.   II. Title.
TK5103.59.D56 2015
621.382′7–dc23   2014034786

ISBN 978-1-107-04942-0 Hardback
ISBN 978-1-107-62753-6 Paperback

Contents

Acronyms

3D	3-dimensional
4G	4th generation
AC	alternating current
ACI	adjacent channel interference
ACO-OFDM	asymmetrically clipped optical orthogonal frequency division multiplexing
ADC	analog-to-digital converter
AGC	automatic gain control
AMC	adaptive modulation and coding
AP	access point
AWGN	additive white Gaussian noise
AZ	azimuth
BB	busy burst
BER	bit-error ratio
BPSK	binary phase shift keying
BRDF	bi-directional reflectance distribution function
CAD	computer-aided design
CCDF	complementary cumulative distribution function
CCI	co-channel interference
CDMA	code division multiple access
CESAR	cellular slot access and reservation
CLT	central limit theorem
CP	cyclic prefix
CSMA/CD	carrier sense multiple access with collision detection
DAC	digital-to-analog converter
DC	direct current
DCO-OFDM	direct-current-biased optical orthogonal frequency division multiplexing
DFE	decision-feedback equalizer
DMT	discrete multi-tone
DSL	digital subscriber line
DSP	digital signal processor
E/O	electrical-to-optical

EL	elevation
FDD	frequency division duplexing
FDMA	frequency division multiple access
FEC	forward error correction
FFE	feed-forward equalizer
FFT	fast Fourier transform
FOV	field of view
FSO	free-space optical
HPA	high-power amplifier
ICI	inter-carrier interference
IEEE	Institute of Electrical and Electronics Engineers
IFFT	inverse fast Fourier transform
IM/DD	intensity modulation and direct detection
IP	Internet protocol
IR	infrared
IrDA	Infrared Data Association
ISI	inter-symbol interference
LDPC	low density parity check
LED	light emitting diode
Li-Fi	light fidelity
LOS	line-of-sight
LTE	long term evolution
M-PAM	multi-level pulse amplitude modulation
M-PAPM	multi-level pulse amplitude and position modulation
M-PPM	multi-level pulse position modulation
M-QAM	multi-level quadrature amplitude modulation
MAC	medium access control
Mbps	megabits per second
MCRT	Monte Carlo ray-tracing
MIMO	multiple-input–multiple-output
MLSD	maximum likelihood sequence detection
MMSE	minimum mean squared error
MRC	maximum ratio combining
NIR	near infrared
NLOS	non-line-of-sight
O/E	optical-to-electrical
O-OFDM	optical orthogonal frequency division multiplexing
OFDM	orthogonal frequency division multiplexing
OFDM-TDMA	OFDM time division multiple access
OFDMA	orthogonal frequency division multiple access
OFDMA-TDD	OFDMA time division duplexing
OOK	on–off keying
OSTBC	orthogonal space-time block codes
OWC	optical wireless communication

P/S	parallel-to-serial
PAM-DMT	pulse amplitude modulation discrete multi-tone
PAPR	peak-to-average-power ratio
PD	photodiode
PDF	probability density function
PDU	protocol data unit
PEP	pairwise error probability
PIM	pulse interval modulation
PIN	positive–intrinsic–negative
PLC	power line communication
PoE	power-over-Ethernet
PSD	power spectral density
PSU	passenger service unit
PWM	pulse width modulation
QoS	quality of service
RC	repetition coding
RF	radio frequency
RGB	red, green, and blue
RMS	root mean square
Rx	receiver
S/P	serial-to-parallel
SE	spectral efficiency
SER	symbol-error rate
SFO-OFDM	spectrally factorized O-OFDM
SIMO	single-input–multiple-output
SINR	signal-to-interference-and-noise ratio
SIR	signal-to-interference ratio
SISO	single-input–single-output
SM	spatial modulation
SMP	spatial multiplexing
SNR	signal-to-noise ratio
TDD	time division duplexing
TDMA	time division multiple access
TIA	transimpedance amplifier
Tx	transmitter
UE	user equipment
U-OFDM	unipolar orthogonal frequency division multiplexing
VLC	visible light communication
VLCC	Visible Light Communications Consortium
WDD	wavelength division duplexing
WDM	wavelength division multiplexing
WDMA	wavelength division multiple access
Wi-Fi	wireless fidelity
WLAN	wireless local area network
ZF	zero forcing

Notation

$*$	linear convolution operator
$\star$	linear discrete convolution operator
$[\cdot]^T$	transpose operator
$\|\cdot\|_F$	Frobenius norm
$\lfloor\cdot\rfloor$	floor operator
$\lceil\cdot\rceil$	ceiling operator
$\mathrm{mod}(\cdot,\cdot)$	modulus of a congruence
a	variable related to the RMS delay spread of the channel, D
A	photosensitive area of the PD
A_0	reference area of $1\ \mathrm{m}^2$
b	bit sequence, reservation indicator
$\mathbf{b}$	bit loading vector in OFDM
B	signal bandwidth
B_c	coherence bandwidth of the optical wireless channel
BER	generalized BER of M-PAM and M-QAM O-OFDM
$\mathrm{BER_{PAM}}$	BER of M-PAM
$\mathrm{BER_{RC}}$	BER of RC
$\mathrm{BER_{SM}}$	BER of SM
$\mathrm{BER_{SMP}}$	BER of SMP
BOTTOM	shifted bottom clipping level
c	speed of light
$c(t)$	chip
$\mathbf{c}$	chip vector
C	mutual information/information rate
$\mathrm{Cov}\,[\cdot]$	covariance operator
d	distance between the transmitter and the receiver on the direct path
d_1	distance between the transmitter and the reflective surface
d_2	distance between the receiver and the reflective surface
$d_H(\cdot,\cdot)$	Hamming distance of two bit sequences
d_{ref}	reference distance
d_s	distance between an intended symbol and the closest interfering symbol

d_{tot}	total distance (including the reflections) a ray travels
d_{Tx}	spacing of the individual transmitters in the optical array
D	RMS delay spread of the channel
$\mathrm{E}[\cdot]$	expectation operator
$E(d_{\text{ref}})$	irradiance at a reference distance d_{ref}
$E_{\text{b(elec)}}$	average electrical bit energy
E_{eye}	irradiance of the eye
$E_{\text{s(elec)}}$	average electrical symbol energy
f	frequency variable
$\mathbf{f}$	OFDM frame vector
$\tilde{\mathbf{f}}$	distorted replica of the OFDM frame vector at the receiver
$\mathbf{f}_{\text{info}}$	vector with the information-carrying subcarriers
$\tilde{\mathbf{f}}_{\text{info}}$	distorted replica of the vector with the information-carrying subcarriers
$F(\cdot)$	non-linear transfer function of the transmitter
F_{OE}	O/E conversion factor for the intended signal
$F_{\text{OE,I}}$	O/E conversion factor for the interfering signal
$F_{\text{O,S}}$	factor for the useful optical symbol power in the intended signal
$g_{\text{h(elec)}}$	electrical path gain
$g_{\text{h(opt)}}$	optical path gain
g_{I}	optical path gain between an interfering transmitter and the receiver
g_{S}	optical path gain between the intended transmitter and the receiver
G_{B}	bandwidth utilization factor
G_{DC}	DC-bias gain
G_{EQ}	equalizer gain
G_{GC}	Gray coding gain
G_{OC}	gain of the optical concentrator
G_{T}	utilization factor for the information-carrying time
G_{TIA}	gain of the TIA
$\mathbf{h}$	impulse response vector of the optical wireless channel
$h(t)$	impulse response of the optical wireless channel
$h_{\text{norm}}(t)$	normalized impulse response of the optical wireless channel
$h_{n_r n_t}$	optical channel gain between transmitter n_t and receiver n_r
$\mathbf{H}$	optical channel matrix in a MIMO setup
$\mathbf{H}_{d_{\text{Tx}}}$	optical channel matrix for a given spacing of transmitters
$\widehat{\mathbf{H}}_{d_{\text{Tx}}}$	optical channel matrix for a given spacing of transmitters with induced link blockage
$H(f)$	Fourier transform of $h(t)$
$H_{\text{norm}}(f)$	Fourier transform of $h_{\text{norm}}(t)$
i	imaginary unit $i = \sqrt{-1}$, index of a direct ray, index of an interfering transmitter
I_{f}	forward current

I_{in}	input current
I_{max}	maximum forward current
$I_{max,norm}$	normalized maximum forward current
I_{min}	minimum forward current
$I_{min,norm}$	normalized minimum forward current
I_{out}	output current
j	index of a polynomial function $\psi(\cdot)$, index of a reflecting ray
$\hat{j}$	index of a normalized clipping level, λ
J	number of normalized clipping levels
k	time-domain sample index in OFDM, index of a ray reflection
k_B	Boltzmann's constant
k_c	chunk time index
K	attenuation factor of the non-linear distortion for the information-carrying subcarriers
K_C	size of coordination cluster of adjacent APs
l	symbol index
L	number of symbols per transmission vector
m	subcarrier index
m_c	chunk frequency index
M	modulation order
M_{SE}	modulation and coding scheme
$\hat{M}_{SE}$	modulation scheme associated with an *a priori* SINR estimate
$\bar{M}_{SE}$	modulation scheme of higher order to be used in the next frame
n	integer polynomial order
n_{os}	number of OFDM symbols per chunk
n_r	index of a receiver/PD in the optical array
n_{sc}	number of subcarriers per chunk
n_{spec}	Lambertian mode number of the specular reflection
n_t	index of a transmitter/LED in the optical array
n_{Tx}	Lambertian mode number of the transmitter
N	number of subcarriers, FFT size
N_0	power spectral density of the AWGN
N_A	number of APs
N_C	number of chunks
N_{CP}	number of CP samples in OFDM
N_{LOS}	number of rays directly impinging on the PD
N_{NLOS}	number of rays which undergo one or multiple reflections
N_r	number of receivers/PDs in the optical front-end
N_{rays}	number of rays
N_{refl}	number of reflections
N_s	average number of neighboring symbols
N_t	number of transmitters/LEDs in the optical front-end
p	scaling factor for the current/optical power levels in M-PAM
$\mathbf{p}$	power loading vector in OFDM

$p_S(s)$	PDF of the clipped OFDM symbol
$p_{\hat{S}}(\hat{s})$	PDF of the unfolded clipped ACO-OFDM symbol
$p_{\mathbf{y}}$	probability density function of the received signal $\mathbf{y}$
$P_{\text{avg,norm}}$	normalized average optical power constraint
$P_{\text{bb,elec}}$	electrical power of the BB signal at a receiver
$P_{\text{b(elec)}}$	average electrical bit power
P_{bg}	optical power of the background illumination
P_{elec}	dissipated electrical power
P_{eye}	optical power irradiating the eye pupil
$P_{\text{I,elec}}$	total interference electrical power
$P_{\text{I,elec,th}}$	threshold interference electrical power
$P_{\text{I,opt}}$	optical power of an interfering transmitter
P_{LOS}	received optical power from the direct rays
$P_{\text{max,norm}}$	normalized maximum optical power constraint
$P_{\text{min,norm}}$	normalized minimum optical power constraint
$P_{\text{N,elec}}$	electrical noise power
P_{NLOS}	received optical power from the reflecting rays
P_{opt}	radiated optical power
$P_{\text{opt}}^{\text{bb}}$	radiated optical power of the BB signal
$P_{\text{opt}}^{\text{PAM}}$	optical power level of an M-PAM symbol for an infinite non-negative linear dynamic range
$P_{\text{opt}}^{\text{SM}}$	optical power level in SM for an infinite non-negative linear dynamic range
P_R	optical power at receiver
$P_{\text{S,elec}}$	intended electrical symbol power
$P_{\text{S,opt}}$	optical power of the intended transmitter
$P_{\text{s(elec)}}$	average electrical symbol power
$P_{\text{s(opt)}}$	average optical symbol power
$P_{\tilde{\text{s}}(\text{opt})}$	effective received optical symbol power in a MIMO setup
$\tilde{P}_{\text{s(opt)},n_{\text{t}}}$	average optical power assigned to transmitter n_{t} in a MIMO setup with power imbalance
PEP_{SM}	PEP of SM
PEP_{SMP}	PEP of SMP
$\text{PL}(d)$	path loss at distance d
P_{T}	optical power of the transmitter
q	elementary electric charge, i.e. $q = 1.6 \times 10^{-19}$ C
$Q(\cdot)$	CCDF of a standard normal distribution
r	radius in a spherical coordinate system
r_{eye}	radius of the eye pupil
r_{FOV}	FOV radius at distance of 20 cm
R	coding rate
R_{b}	bit rate
R_{load}	load resistance

$R_{\text{LED}}(\theta, n_{\text{Tx}})$	generalized Lambertian radiation pattern of the LED
R_{s}	symbol rate
R_{th}	user reservation threshold
s	user score for fair resource reservation
$s(t)$	symbol
$\hat{s}(t)$	unfolded ACO-OFDM symbol
$\bar{s}(t)$	unfolded and debiased ACO-OFDM symbol
$\mathbf{s}$	symbol vector
$\tilde{\mathbf{s}}$	distorted replica of the symbol vector at the receiver
S	number of subbands of subcarriers, number of groups of cells
S_{PD}	responsivity of the PD
SE	spectral efficiency of the modulation schemes for OWC
SE_M	spectral efficiency of a modulation and coding scheme
SINR	electrical SINR in O-OFDM
SNR	electrical SNR in O-OFDM
SNR_{Rx}	electrical SNR at the receiver side in a MIMO setup
SNR_{Tx}	electrical SNR at the transmitter side in a MIMO setup
t	time variable
T	absolute temperature
T_{s}	symbol duration in PPM and PAM
T_{OF}	transmittance of the optical filter
TOP	shifted top clipping level
u	dummy integration variable
U	number of users
U_{A}	number of users served by an AP
$U(t)$	unit step function
$v(t)$	impulse response of the pulse shaping filter
$V(f)$	Fourier transform of $v(f)$
V_{f}	forward voltage
$w(t)$	AWGN at the receiver
$\mathbf{w}$	AWGN vector at the receiver
$w_{\text{clip}}(t)$	uncorrelated non-Gaussian time-domain non-linear distortion noise
$\mathbf{W}$	AWGN vector at the frequency domain subcarriers
$\mathbf{W}_{\text{clip}}$	additive Gaussian non-linear distortion noise at the information-carrying subcarriers
$x(t)$	biased information-carrying signal
$\mathbf{x}$	biased information-carrying signal vector
$\hat{\mathbf{x}}$	decoded signal vector at the receiver
x_{Rx}	position offset of the receiver array on the X-axis
X	direction in a Cartesian coordinate system
$y(t)$	received signal
$\mathbf{y}$	received signal vector
y_{Rx}	position offset of the receiver array on the Y-axis

Y	direction in a Cartesian coordinate system
z	counter in summation of $W(f)$
Z	direction in a Cartesian coordinate system
$Z_{\mathbf{x}}$	length of the biased information-carrying signal vector
$Z_{\mathbf{h}}$	length of the impulse response vector of the optical wireless channel
α	scaling factor for the signal power
α_A	AP associated with UE μ_U
β_A	AP associated with UE ν_U
β_{DC}	DC-bias current
γ	electrical SINR at receiver
$\hat{\gamma}$	*a priori* estimate of the electrical SINR at receiver
$\gamma_{b(elec)}$	undistorted electrical SNR per bit at the transmitter
Γ	electrical SINR target
$\Gamma_{b(elec)}$	effective electrical SNR per bit at the receiver
Γ_{min}	minimum electrical SINR target
δ	optical power imbalance factor between the individual transmitters
$\delta(t)$	Dirac delta function
$\Delta_{d_{Tx}}^{SNR_{Rx}}$	penalty on the received electrical SNR for a given spacing of transmitters
$\widehat{\Delta}_{d_{Tx}}^{SNR_{Rx}}$	penalty on the received electrical SNR for a given spacing of transmitters with induced link blockage
ϵ	user indicator for access of an idle chunk
ζ	shadowing component, index of user
η	fraction of the light reflected in a diffuse Lambertian fashion
θ	zenith angle in a spherical coordinate system
θ_{eye}	angle subtended by the eye pupil and the origin at $d = 20$ cm
$\theta_{FOV,Rx}$	FOV semi-angle of receiver
$\theta_{FOV,Tx}$	FOV semi-angle of transmitter
θ_{inc}	incident angle of the incoming light/ray at the reflective surface
θ_{obs}	observation angle of the outgoing light/ray from the reflective surface
θ_{Rx}	incident angle of the receiver from the reflective surface
$\theta_{Rx,d}$	incident angle of the receiver on the direct path
θ_{Tx}	observation angle of the transmitter towards the reflective surface
$\theta_{Tx,d}$	observation angle of the transmitter on the direct path
κ	factor of standard deviations quantifying the DC bias in DCO-OFDM
λ	normalized clipping level, wavelength
λ_{bottom}	normalized bottom clipping level

λ_{top}	normalized top clipping level
$\Lambda(f)$	variable related to $V(f)$ and $H(f)$ by (2.20)
μ	mean of the clipped DCO-OFDM symbol
μ_{U}	UE associated with AP α_{A}
ν_{U}	UE associated with AP β_{A}
ξ	path loss exponent
$\Xi(\cdot)$	piecewise polynomial transfer function of the transmitter
π	number pi, $\pi \approx 3.14$
ρ	reflection coefficient of the reflective surface
σ	standard deviation of the OFDM time-domain signal
σ_{AWGN}	standard deviation of the AWGN
σ_{clip}	standard deviation of the non-linear distortion noise
σ_{shad}	standard deviation of log-normal shadowing
τ	dummy integration variable
ϕ	azimuth angle in a spherical coordinate system
$\phi(\cdot)$	PDF of a standard normal distribution
$\Phi(\cdot)$	linearized transfer function of the transmitter denoting the double-sided signal clipping
$\psi(\cdot)$	polynomial function of non-negative integer order, n
Ψ	user priority penalty factor
$\Psi(\cdot)$	normalized non-linear transfer function of the transmitter
$\hat{\Psi}(\cdot)$	unfolded normalized non-linear transfer function in ACO-OFDM
$\mathcal{A}$	set of chunks assigned to a user
$\#\mathcal{A}$	number of chunks assigned to a user
$\mathcal{G}$	group of APs/cells
$\mathcal{G}_{\beta_{\text{A}}}$	group of APs β_{A}
$\mathcal{I}$	integral structure for calculation of the non-linear distortion parameters
$\mathcal{M}$	set of supported modulation schemes
$\#\mathcal{M}$	cardinality of $\mathcal{M}$
$\mathcal{N}(\mu, \sigma^2)$	normal distribution with mean μ and variance σ^2 of the unclipped OFDM symbol
$\mathcal{R}(\theta, \phi)$	BRDF
$\mathcal{R}_{\text{in}}(\theta, \phi, \lambda)$	portion of the BRDF related to the incoming light

1 Introduction

Optical communication is any form of telecommunication that uses light as the transmission medium. Having originated in ancient times in the form of beacon fires and smoke signals that convey a message, optical wireless communication (OWC) has evolved to a high-capacity complementary technology to radio frequency (RF) communication. OWC systems utilize wavelengths in the infrared (IR) spectrum for IR communication and the visible light spectrum for visible light communication (VLC). Because of the availability of a huge license-free spectrum of approximately 670 THz, OWC has the potential to provide wireless links with very high data rates. In this book, optical modulation schemes, as well as signal processing and networking techniques, are presented to maximize the throughput of optical wireless networks using off-the-shelf components.

1.1 History of OWC

Examples of OWC in the form of beacon fires and smoke signals to convey a message can be found in almost all cultures. Semaphore lines are the earliest form of technological application of OWC [1]. The French engineer Claude Chappe built the first optical telegraph network in 1792. His semaphore towers enabled the transmission of 196 information symbols encoded in the position of two arms connected by a crossbar. As another example of early OWC, the heliograph is a wireless solar telegraph that signals flashes of sunlight by pivoting a mirror or interrupting the beam with a shutter. After the invention of the Morse code in 1836, navy ships communicated by means of a signal lamp with on-shore lighthouses for navigation. In 1880, Alexander Graham Bell demonstrated the first implementation of a free-space optical (FSO) link in the form of the photophone [2]. By using a vibrating mirror at the transmitter and a crystalline selenium cells at the focal point of a parabolic receiver, Bell was able to modulate a voice message onto a light signal.

The recent advancements in OWC technology gained significant pace after the pioneering work of Gfeller and Bapst in 1979 [3]. They showed the potential of OWC for high-capacity in-house networks promising hundreds of THz bandwidth of electromagnetic spectrum in the optical domain. One branch of OWC is targeted at outdoor FSO links over long distances which are generally realized through highly directional laser diodes as transmitters [4]. At the receiver side, generally, a photodiode (PD) is employed. Another branch of OWC focuses on indoor mobile wireless networks, and it is realized through diffuse light emitting diodes (LEDs) as transmitters [3]. The first indoor OWC system was reported by Gfeller and Bapst in 1979 [3]. At a center wavelength of 950 nm in the IR spectrum, the system was capable of achieving 1 Mbps using on–off keying (OOK) modulation and diffuse radiation for a coverage of an office room. In 1996, Marsh and Kahn demonstrated an indoor diffuse OOK IR system with a data rate of 50 Mbps [5]. Later in 2000, Carruthers and Kahn presented a faster OOK IR system implementation with a data rate 70 Mbps and a potential of up to 100 Mbps [6]. Tanaka *et al.* first considered white LEDs to convey information in addition to serving the primary functionality of illumination in an indoor setup. In 2003, they presented a OOK VLC system setup with up to 400 Mbps data rate [7]. Afgani *et al.* [8, TridentCom, 2006] showed for the first time, using a proof-of-concept demonstrator, that the high crest factor in orthogonal frequency division multiplexing (OFDM), typically a disadvantage in RF communication, can be turned into an advantage for intensity modulation and direct detection (IM/DD). They implemented the direct-current-biased optical OFDM (DCO-OFDM) transmission scheme, which was later used by other research groups. Vucic *et al.* ascertained the potential of VLC systems with a demonstration of a 500 Mbps data rate [9]. Their implementation was based on DCO-OFDM with bit and power loading and symmetric signal clipping. By separate modulation of the red, green, and blue (RGB) modes of an RGB white LED in a wavelength division multiplexing (WDM) fashion and by employing respective optical filters at the receiver, they have also been able to demonstrate 800 Mbps of a single RGB LED luminary [10]. By the use of a similar DCO-OFDM setup with a larger modulation bandwidth, Khalid *et al.* presented a link implementation that can achieve 1 Gbps with a single phosphor-coated white LED [11]. Later, they also demonstrated a 3.4 Gbps link with an off-the-shelf RGB LED [12]. Another similar gigabit/s OWC system with phosphor-coated white LED has been demonstrated by Azhar *et al.* using a 4×4 multiple-input–multiple-output (MIMO) configuration [13]. Recently, Tsonev *et al.* reported a data rate of 3.5 Gbps from a single-color micro LED in a single-input–single-output (SISO) setting [14, 15]. The fundamentals of optical modulation and signal processing that enable the aforementioned data rates are presented in this book alongside state-of-the-art networking concepts [16–18]. These developments promote OWC to an emerging wireless networking technology – light fidelity (Li-Fi), a term coined by Harald Haas at TEDGlobal in 2011 [19].

Originally targeted at the near infrared (NIR) spectrum [3, 20–22], the optical wireless link was meant for short-range communications. Since 1993, a standardized set of protocols of the Infrared Data Association (IrDA) [23] have been implemented for wireless infrared communication in portable devices, such as mobile phones, laptops, cameras, remote controls, and many more. With the advancements of solid-state lighting technology in recent years, LEDs are replacing incandescent light bulbs because of their reliability and higher energy efficiency, e.g. 5% vs. 30% in favor of LEDs [24]. In addition to illumination, LEDs are also envisioned to provide high-capacity wireless data broadcast [25–33]. Standardization of VLC research is strongly supported by the Visible Light Communications Consortium (VLCC) in Japan [34]. In 2011, the Institute of Electrical and Electronics Engineers (IEEE) published a standard for VLC, IEEE 802.15.7 − 2011, "IEEE Standard for Local and Metropolitan Area Networks, Part 15.7: Short-Range Wireless Optical Communication Using Visible Light" [35].

1.2   Advantages of OWC

In the last two decades, unprecedented spread of wireless communication systems has been witnessed. While at the beginning these systems were only able to provide voice service and some rudimentary data services, they have now matured to high-speed packet data networks which allow Internet browsing at the same speed as is achieved with fixed line connections [36]. However, there is still the need to increase data throughput and, consequently, data rates [37].

With the increasing popularity of smartphones, the wireless data traffic of mobile devices is growing exponentially. There have been many independent warnings of a looming "RF spectrum crisis" [38] as mobile data demands continue to increase, while the network spectral efficiency saturates despite newly introduced standards and great technological advancements in the field. By 2015, it is expected that total wireless data traffic will reach 6 exabytes per month, potentially creating a 97% gap between the traffic demand per device and the available data rate per device in the mobile networks [39]. In addition, it is estimated that by 2017 more than 11 exabytes of data traffic will have to be transferred through mobile networks every month [40]. Recently, the Wireless Gigabit Alliance has proposed the utilization of the mm-waves in the license-free 60 GHz band, where the availability of 7 GHz bandwidth enables 7 Gbps short-range wireless links [41]. The 60 GHz band has also been considered as a part of the IEEE 802.11ad framework for very high throughput data links in wireless local area networks (WLANs) using MIMO techniques [42]. However, due to the high path loss of the radio waves in this spectrum range, 60 GHz links are highly directional, and, therefore, require sophisticated digital beamforming and tracking algorithms for application in mobile wireless networks.

Since the RF spectrum is limited and expensive, new and complementary wireless transmission techniques are currently being explored that can relieve the spectrum utilization. One such promising emerging alternative approach is OWC, which offers many advantages over RF transmission. Most recently, VLC has been identified as a potential solution for mitigating the looming RF spectrum crisis. VLC is particularly enticing as lighting is a commodity that has been integrated into virtually every inhabited environment, and sophisticated infrastructures already exist. The use of the visible light spectrum for high-speed data communication is enabled by the emergence of the LED, which at the same time is at the heart of the next wave of energy-efficient illumination. In that sense, the concept of combining the functions of illumination and communication offers the potential for tremendous cost savings and carbon footprint reductions. First, the deployment of VLC access points (AP) becomes straightforward as the existing lighting infrastructure can be reused. Off-the-shelf technologies, such as power line communication (PLC) and power-over-Ethernet (PoE), are viable backhaul solutions for retrofit installations and new installations, respectively. Second, because lighting is on most of the time in indoor environments even during day time, the energy used for communication is significantly reduced as a result of the piggybacking of data on illumination. However, even if illumination is not required, energy-efficient IM/DD techniques exist that allow data communication, even if the lights are visually off [43]. These are already compelling benefits, but the case does not end there. In OWC, the signal can occupy license-free wavelengths in the visible light spectrum from 380 nm to 750 nm, and/or the NIR spectrum from 750 nm to 2.5 μm. The total available bandwidth resource amounts to approximately 670 THz, which is a factor of 10, 000 larger than the RF spectrum including the 60 GHz band. In addition to being a complementary non-interfering solution alongside the RF technology, OWC has the advantage of license-free operation over a huge spectrum resource. In addition, very high data rates can be realized by the use of low-cost front-ends with commercially available LEDs and PDs [20]. Furthermore, it is free of any health concerns as long as eye safety regulations are fulfilled [44]. This constraint is much less severe when using incoherent LEDs rather than laser diodes. With the advent of highly efficient high-power incoherent LEDs and highly sensitive PDs, OWC has become a viable candidate for medium-range indoor data transmission that can contribute to the cause of solving the spectrum deficit.

1.3   Application areas

OWC is generally realized in a line-of-sight (LOS) or a non-line-of-sight (NLOS) communication setup [20, 21]. LOS links can be generally employed in static communication scenarios such as indoor sensor networks, where a fixed position and alignment between the transmitter and receiver are maintained. In mobile environments such as commercial offices, mechanical or electronic beam steering [4] can be used to maintain an LOS connection. Such techniques, however, increase the cost of the optical front-ends. Therefore, in a mobile OWC network, where LOS links are likely to be blocked, transmission robustness can be facilitated through NLOS communication. Single-carrier pulse modulation techniques such as pulse width modulation (PWM), pulse interval modulation (PIM), pulse position modulation (PPM), and pulse amplitude modulation (PAM) experience inter-symbol interference (ISI) in the dispersive NLOS channel, and they therefore exhibit limited data rates unless computationally expensive equalizers are used [4, 20, 45]. Because of its inherent robustness to multipath fading, OFDM with multi-level quadrature amplitude modulation (M-QAM) is envisaged to enable NLOS communication, and therefore high-capacity wireless networking [8, 46, 47].

In addition, due to the fact that light does not propagate through opaque objects and walls, optical wireless signals can be confined within a room. This feature inherently eliminates concerns over the intercepting and eavesdropping of the transmission, resulting in secure indoor data links and networks. The same feature can be exploited to eliminate interference between neighboring cells. Furthermore, OWC is free of any health concerns as long as eye safety regulations are fulfilled [44]. Since optical radiation does not interfere with other electromagnetic waves or with the operation of sensitive electronic equipment, OWC enables safe data transmission in areas where RF communication and electromagnetic radiation are prohibited or refrained to avoid interference with critical systems. These include aviation, homeland security, hospitals, and healthcare, as well as petrochemical and nuclear power plants. Last but not least, radio waves are strongly attenuated in water, disallowing underwater RF transmission. However, since light propagates through water, OWC can be employed for underwater communication.

1.4 Li-Fi

To date, research in the field of OWC has been focused on successful implementations of physical-link connections and proofs of the concept [31]. For the realization of a mobile communication system, however, a full networking solution is required. This is what is referred to as Li-Fi: the networked, mobile, high-speed OWC solution [48]. The vision is that a Li-Fi wireless network would complement existing heterogeneous RF wireless networks, and would provide significant spectrum relief by allowing cellular and wireless fidelity (Wi-Fi) systems to off-load a significant portion of wireless data traffic.

1.4.1 Modulation

A seamless all-optical wireless network would require ubiquitous coverage provided by the optical front-end elements. This necessitates the usage of a large amount of Li-Fi enabled lighting units. The most likely candidates for front-end devices in VLC are incoherent LEDs for solid-state lighting because of their low cost. Due to the physical properties of these components, information can only be encoded in the intensity of the emitted light. As a result, VLC can be realized as an IM/DD system, which means that the modulation signal has to be both real-valued and unipolar non-negative. This limits the application of the well-researched and developed modulation schemes from the field of RF communications. Techniques such as PWM, PPM, OOK, and PAM can be applied in a relatively straightforward fashion. As the modulation speeds are increased, however, these particular modulation schemes begin to suffer from the undesired effects of ISI due to the frequency selective optical wireless channel. Hence, a more resilient technique such as OFDM is required. OFDM allows for adaptive bit and power loading of different frequency subbands according to the communication channel properties [49, 50]. This leads to optimal utilization of the available resources. Such channel conditions are introduced by the frequency response of an off-the-shelf LED which has a maximum 3-dB modulation bandwidth of a few tens of MHz [11, 12]. Further benefits of this modulation scheme include simple equalization with single-tap equalizers in the frequency domain, as well as the ability to avoid low-frequency distortion caused by flickering background radiation and the DC-wander effect in electrical circuits.

Conventional OFDM signals are complex-valued and bipolar in nature. Therefore, the standard RF OFDM technique has to be modified in order to become suitable for IM/DD systems. A straightforward way to obtain a real-valued OFDM signal is to impose a Hermitian symmetry constraint on the subcarriers in the frequency domain. However, the resulting time-domain signal is still bipolar. One way of obtaining a unipolar signal is to introduce a positive DC bias. The resulting unipolar modulation scheme is known as DCO-OFDM. The addition of the constant biasing level leads to a significant increase in electrical energy consumption. However, if the light sources are used for illumination at the same time, the light output as a result of the DC bias is not wasted as it is used to fulfill the illumination function. Only if illumination is not required, such as in the uplink of a Li-Fi system, the DC bias can significantly compromise energy efficiency. Therefore, researchers have devoted significant efforts to designing an OFDM-based modulation scheme which is purely unipolar. Some well-known solutions include: asymmetrically clipped optical OFDM (ACO-OFDM) [51], PAM discrete multi-tone (PAM-DMT) [52], flip-OFDM [53], unipolar OFDM (U-OFDM) [54], and spectrally factorized optical OFDM (SFO-OFDM) [55]. The general disadvantage of all these techniques is a 50% loss in spectral efficiency and data rates.

From a networking perspective, OFDM offers a straightforward multiple access implementation as subcarriers can be allocated to different users resulting in orthogonal frequency division multiple access (OFDMA). The merits of OFDM have already been recognized, and it is used in IEEE 802.11 Wi-Fi systems. Also, OFDMA is used in the 4th generation (4G) long-term evolution (LTE) standard for cellular mobile communications. Therefore, the application of OFDM in optical mobile networks would allow the use of the already established higher level communication protocols used in IEEE 802.11 and LTE.

1.4.2   Multiple access

A seamless all-optical networking solution can only be realized with a suitable multiple access scheme that allows multiple users to share the communication resources without any mutual cross-talk. Multiple access schemes used in RF communications can be adapted for OWC as long as the necessary modifications related to the IM/DD nature of the modulation signals are performed. OFDM comes with a natural extension for multiple access – OFDMA. Single-carrier modulation schemes such as PPM and PAM require an additional multiple access technique such as frequency division multiple access (FDMA), time division multiple access (TDMA), or code division multiple access (CDMA).

OFDMA has been compared with TDMA and CDMA in terms of the electrical power requirement in a flat fading channel with additive white Gaussian noise (AWGN) and a positive infinite linear dynamic range of the transmitter [56]. FDMA has not been considered due to its close similarity to OFDMA, and the fact that OWC does not use superheterodyning. In addition, due to the limited modulation bandwidth of the front-end elements, FDMA would not present an efficient use of the LED modulation bandwidth. CDMA demonstrates the highest electrical power requirement, since the use of unipolar signals creates significant ISI. TDMA is shown to marginally outperform OFDMA in this setup. The increased power requirement of OFDMA comes from the higher DC-bias level needed to condition the OFDM signal within the positive dynamic range of the LED. However, in a practical VLC scenario, where the functions of communication and illumination are combined, the difference in power consumption between OFDMA and TDMA would diminish as the excess DC-bias power would be used for illumination purposes. Furthermore, TDMA and CDMA systems experience low-frequency distortion noise due to DC wander in electrical components or flickering of background illumination sources, as well as severe ISI in the practical dispersive and frequency selective channel. Therefore, the design complexity of TDMA and CDMA systems increases as suitable techniques to deal with these issues need to be implemented.

In OWC, there exists an additional alternative dimension for achieving multiple access. This is the color of the LED, and the corresponding technique is wavelength division multiple access (WDMA). This scheme can reduce the complexity of signal processing at the expense of increased hardware complexity. This is because each AP would require multiple LEDs and PDs with narrow-band emission and detection spectra. Alternatively, narrow-band optical filters can be employed. However, the variation of the center wavelength generally results in variation of the modulation bandwidth, the optical emission efficiency of the LED, and the responsivity of the PD. This corresponds to a variation of the signal-to-noise ratio (SNR) and capacity in the different multiple access channels, which complicates the fair distribution of communication resources to multiple users.

1.4.3   Uplink

Until now, research has primarily focused on maximizing the transmission speeds over a single unidirectional link [11–13]. However, for a complete Li-Fi communication system, full duplex communication is required, i.e. an uplink connection from the mobile terminals to the optical AP has to be provided. Existing duplex techniques used in RF such as time division duplexing (TDD) and frequency division duplexing (FDD) can be considered, where the downlink and the uplink are separated by different time slots or different frequency bands, respectively. However, FDD is more difficult to realize due to the limited bandwidth of the front-end devices, and because superheterodyning is not used in IM/DD systems. TDD provides a viable option, but imposes precise timing and synchronization constraints similar to the ones needed for data decoding. However, TDD assumes that both the uplink and the downlink transmissions are performed over the same physical wavelength. This can often be impractical as visible light emitted by the user terminal may not be desirable [57]. Therefore, the most suitable duplex technique in Li-Fi is wavelength division duplexing (WDD), where the two communication channels are established over different electromagnetic wavelengths. Using IR transmission is one viable option for establishing an uplink communication channel [57]. A first commercially available full duplex Li-Fi modem using IR light for the uplink channel has recently been announced by pureLiFi [58]. There is also the option to use RF communication for the uplink [57]. In this configuration, Li-Fi can be used to off-load a large portion of data traffic from the RF network, thereby providing significant RF spectrum relief. This is particularly relevant since there is a traffic imbalance in favor of the downlink in current wireless communication systems.

1.4.4   The attocell

In the past, wireless cellular communication has significantly benefited from reducing the inter-site distance of cellular base stations. By reducing the cell size, network spectral efficiency has been increased by two orders of magnitude in the last 25 years. More recently, different cell layers composed of microcells, picocells, and femtocells have been introduced. These networks are referred to as heterogeneous networks [59, 60]. Femtocells are short-range, low transmission power, low-cost, plug-and-play base stations that are targeted at indoor deployment in order to enhance coverage. They use either cable Internet or broadband digital subscriber line (DSL) to backhaul to the core network of the operator. The deployment of femtocells increases the frequency reuse, and hence throughput per unit area within the system, since they usually share the same bandwidth with the macrocellular network. However, the uncoordinated and random deployment of small cells also causes additional inter- and intra-cell interference which imposes a limit on how dense these small base stations can be deployed before interference starts offsetting all frequency reuse gains.

The small cell concept, however, can easily be extended to VLC in order to overcome the high interference generated by the close reuse of radio frequency spectrum in heterogeneous networks. The optical AP is referred to as an attocell [61]. Since it operates in the optical spectrum, the optical attocell does not interfere with the macrocellular network. The optical attocell not only improves indoor coverage, but since it does not generate any additional interference, it is able to enhance the capacity of the RF wireless networks.

Li-Fi attocells allow for extremely dense bandwidth reuse due to the inherent properties of light waves. Studies on the deployment of indoor optical attocells are presented in Chapter 2 and Chapter 7 of this book. The coverage of each single attocell is very limited, and walls prevent the system from experiencing co-channel interference (CCI) between rooms. This precipitates the need to deploy multiple APs to cover a given space. However, due to the requirement for illumination indoors, the infrastructure already exists, and this type of cell deployment results in the aforementioned very high, practically interference-free bandwidth reuse. Also a byproduct of this is a reduction in bandwidth dilution over the area of each AP, which leads to an increase in the capacity available per user. The user data rate in attocell networks can be improved by up to three orders of magnitude [62].

Moreover, Li-Fi attocells can be deployed as part of a heterogeneous VLC-RF network. They do not cause any additional interference to RF macro- and picocells, and hence can be deployed within RF macro-, pico-, and even femtocell environments. This allows the system to vertically hand-off users between the RF and Li-Fi subnetworks, which enables both free user mobility and high data throughput. Such a network structure is capable of providing truly ubiquitous wireless network access.

1.4.5   Cellular network

The deployment of multiple Li-Fi attocells provides ubiquitous data coverage in a room in addition to providing nearly uniform illumination. This means that a room contains many attocells forming a very dense cellular attocell network. A network of such density, however, requires methods for intra-room interference mitigation, while there is no inter-room interference if the rooms are separated by solid walls. Interference mitigation techniques used in RF cellular networks such as the busy burst (BB) principle [18], static resource partitioning [16, 17, 63], or fractional frequency reuse [64] have been considered. The unique properties of optical radiation, however, offer specific opportunities for enhanced interference mitigation in optical attocell networks. Particularly important is the inability of light to penetrate solid objects, which allows interference to be managed in a more effective manner than in RF communications. According to [62], for example, the VLC interference mitigation caused by solid objects in a typical indoor environment leads to a tremendous increase in the area spectral efficiency over LTE-based RF femtocell network deployment in the same indoor office environment. The presented results highlight that the improvement with respect to the area spectral efficiency can reach a factor of up to 1000 in certain scenarios.

Essential techniques for increasing wireless system capacity such as beamforming are relatively straightforward to use in VLC as the beamforming characteristic is an inherent, device-specific property related to the field of view (FOV), and no computationally complex algorithms and multiple transmitting elements are required. A simple example is provided with the technique of joint transmission in indoor VLC downlink cellular networks [65]. The application of multiple simple narrow-emission-pattern transmitters at each attocellular AP results in significant CCI reduction. The technique allows the cellular coverage area to be broken down further into areas of low interference and areas that are subject to higher interference, typically at the cell edges. Corresponding frequency allocation and constructively superimposed joint transmission can then be performed to increase the overall throughput distribution over the coverage area. A similar concept can also be realized at the receiver side, where multiple receiver elements with a narrow FOV provide a means for enhanced interference mitigation capabilities. The narrow FOV enables each photodetector to scan only a fraction of the available space. The overall combination of all photodetectors provides a wide FOV. This discretization of the receiver eyesight allows interference to be avoided by careful recombination of the output signals from each receiver element. These are only some examples of the cellular network research that is being conducted in the field of OWC.

1.5   Challenges for OWC

The following challenges are relevant for the implementation of an OWC system in practical single-link and multi-user communication scenarios. First, optical transmitter front-ends based on off-the-shelf LEDs exhibit a strong non-linear transfer of the information-carrying signal. Therefore, the optimum conditioning of the time-domain signal within the limited dynamic range of the transmitter front-end is essential in order to minimize the resulting non-linear signal distortion and to maximize the system throughput. In order to formulate this optimization problem, the mathematical details of the optical-to-electrical (O/E) signal conversion of the unipolar optical signals are required. Since the energy efficiency of the system is measured by the amount of electrical power required for a given quality of service (QoS), a relationship with the output optical power needs to be established. Through pre-distortion of the signal with the inverse of the non-linear transfer function, the dynamic range of the transmitter can be linearized between levels of minimum and maximum radiated optical power. While single-carrier signals can fit within the linearized dynamic range of the transmitter without distortion, in an OFDM system the non-linear distortion for a given signal biasing setup needs to be analyzed. Therefore, the achievable information rates of the OFDM system for a practical linear dynamic range of the transmitter under average electrical power and average optical power constraints are to be established.

Currently, OWC systems with IM/DD cannot fully utilize the entire available optical spectrum, because of the small electrical modulation bandwidth compared to the optical center wavelength of the optical front-ends. Therefore, system designers often resort to increasing the signal bandwidth beyond the 3-dB electrical bandwidth of the optical elements for the sake of increasing the system throughput. However, such an approach requires channel equalization techniques such as linear and non-linear equalization for single-carrier signals, and bit and power loading for multi-carrier signals. This requires channel knowledge at the receiver and the transmitter. As a result, the spectral efficiency and electrical SNR requirement of single-carrier and multi-carrier modulation schemes need to be compared in a flat fading channel and a dispersive channel under an average electrical power constraint and minimum, average, and maximum optical power constraints. Moreover, this comprehensive comparison needs to take into account the equalization penalties and the total invested electrical signal power, i.e. alternating current (AC) power and DC power.

Finally, the system model and the optimum front-end biasing setup are often tailored only to a single-link OWC scenario. Capacity enhancing techniques, where multiple LEDs are employed at the transmitter and multiple PDs are employed at the receiver, are still an open issue. The mechanisms that increase the probability of detection of the individual signals and the associated diversity techniques need to be investigated further in the context of MIMO systems. In addition, studies of the OWC systems are to be expanded with the simulation and optimization of multiple access scenarios in a network of mobile users. Because of the fact that the center wavelength is significantly larger than the modulation bandwidth of the optical front-ends, wavelength reuse in cellular OWC systems can be performed without a perceivable reduction of capacity as opposed to RF cellular systems. Therefore, a larger insight is to be gained into the maximization of the capacity of cellular OWC networks with a transition towards autonomous self-organizing interference-aware networks. This book addresses these challenges in the following chapters.

1.6   Summary

Research in OWC over the past ten years has primarily been focused on finding an optimum modulation scheme for point-to-point VLC links with IM/DD, taking into account that VLC may serve two simultaneous functions: (a) illumination and (b) gigabit/s wireless communication. The predominant sources of signal distortion in such systems are frequency dependent. This constitutes one key reason why there is now a general understanding that OFDM is the most suitable choice as a digital modulation scheme for Li-Fi, and there are good technical reasons to reconsider the IEEE 802.15.7 VLC standard. The straightforward multiple access technique that OFDMA provides at almost no additional complexity and its compatibility to state-of-the art wireless standards such as IEEE 802.11 and LTE further favor the selection of this modulation and multiple access scheme.

The realization of a bi-directional connection also seems to have been addressed successfully to an extent that the first commercial bi-directional point-to-point Li-Fi systems are available. The most practical solution to the uplink channel realization is to consider the IR or RF spectrum. The confidence brought by encouraging recent research results and by the successful VLC link-level demonstrations, has now shifted the focus towards an entire Li-Fi attocell networking solution. The unique physical properties of light promise to deliver very densely packed high-speed network connections resulting in orders of magnitude improved user data rates. Based on these very promising results, it seems that Li-Fi is rapidly emerging as a powerful wireless networking solution to the looming RF spectrum crisis, and an enabling technology for the future Internet-of-Things. Based on past experience that the number of wireless applications increases by the square of the number of available physical connections, Li-Fi could be at the heart of an entire new industry for the next wave of wireless communications.

2 Optical wireless communication

2.1 Introduction

In optical wireless communication (OWC), the light intensity of a light emitting diode (LED) is modulated by a message signal. After propagating through the optical wireless channel, the light message is detected by a photodiode (PD). Key characteristics of the optical transmitter (Tx) and receiver (Rx) include their optical spectral response, electrical modulation bandwidth, radiation/detection patterns, optical power output of the LED, the photosensitive area, and the noise figure of the PD. The optical wireless channel has been shown to be a linear, time-invariant, memoryless system with an impulse response of a finite duration [21]. The primary characteristic of the channel is the path loss. An accurate representation of the light distribution in an indoor setup can be obtained by means of a Monte Carlo ray-tracing (MCRT) simulation. Channel modeling confirms that the path loss in dB is linear over logarithmic distance, and it ranges between 27 dB and 80 dB for line-of-sight (LOS) and non-line-of-sight (NLOS) communication scenarios in the considered setup [16, 66]. At high data rates, where the signal bandwidth exceeds the channel coherence bandwidth, the channel can be characterized as a frequency selective channel due to dispersion [67]. There is no fast fading, but the signal undergoes slow fading. Ray-tracing simulations show that slow fading can be modeled as a log-normally distributed random variable [16]. The delay spread of dispersive optical wireless channels can be accurately modeled by a rapidly decaying exponential impulse response function. Root-mean-squared (RMS) delay spreads between 1.3 ns and 12 ns are reported for LOS links, whereas RMS delay spreads between 7 ns and 13 ns are reported for NLOS links [21]. In order to counter the channel effect, single-carrier pulse modulation techniques, e.g. multi-level pulse position modulation (M-PPM) and multi-level pulse amplitude modulation (M-PAM), employ a linear feed-forward equalizer (FFE) or a non-linear decision-feedback equalizer (DFE) with zero forcing (ZF) or minimum mean-squared error (MMSE) criteria at the expense of increased computational complexity. The equalizers incur a penalty on the signal-to-noise ratio (SNR). In multi-carrier modulation, e.g. optical orthogonal frequency division multiplexing (O-OFDM) with multi-level quadrature amplitude modulation (M-QAM), bit and power loading of the individual subcarriers can be employed in order to minimize the SNR penalty in the frequency selective channel. As a result, low complexity single-tap equalizers can be used. Similar effects arise due to the use of off-the-shelf components with low modulation bandwidth. Therefore, equalization techniques are essential for the realization of high-capacity OWC links.

In this chapter, the basic concepts of an OWC link setup are presented. The geometry of the LOS and NLOS communication scenarios is defined, and the key characteristics of the transmitter and the receiver used for system design and simulation are introduced. The statistical model of the optical wireless channel is presented, including key characteristics such as path loss, delay spread, and coherence bandwidth. The SNR penalties of the common channel equalization techniques are presented. In addition, the mathematical details of ray-tracing techniques such as the deterministic algorithm and the Monte Carlo ray-tracing algorithm are summarized. Ray-tracing is a powerful and flexible tool to calculate the path loss and delay spread in any geometric setup of an OWC system. A case study of a cellular OWC system inside an aircraft cabin is presented. It provides key insights into path loss models for LOS and NLOS communications, including shadowing, as well as the signal-to-interference ratio (SIR) maps with wavelength reuse.

2.2   System setup

The geometry of a wireless communication scenario is defined by the position and radiation/detection characteristics of the transmitters and receivers in an indoor or outdoor environment, with certain reflection properties of the objects in the setup. Based on the propagation path of the light radiated by the transmitter and detected by the receiver, there are two general link arrangements, i.e. LOS and NLOS communications [21]. In addition, a cellular network can be deployed in order to maximize the coverage and capacity over the area of the OWC setup [16, 63, 68]. In this section, the building blocks of the transmitter and receiver front-ends are introduced, and the general communication setup arrangements are discussed.

A generalized OWC link is illustrated in Fig. 2.1. The transmitter consists of a digital signal processor (DSP) with a digital-to-analog converter (DAC), which cater for the modulation of the digital information bits and their transformation into an analog current signal. The current drives the optical emitter, i.e. an LED or an array of LEDs. Here, the information-carrying current signal is transformed into optical intensity. The optical signal can be passed through an optical system in order to further shape the transmitted beam. Here, an optical amplifier lens, a collimator, or a diffusor can be employed to concentrate or broaden the beam. The optical signal is then transmitted over the optical wireless channel. A portion of the optical energy is absorbed by the objects in the environment, and the rest is reflected back in a diffuse or specular fashion [16]. LOS and NLOS signal components arrive at the receiver. An optical filter can be applied to select a portion of interest in the optical spectrum. In addition, the optical filter greatly reduces the interference from ambient light. Thereafter, the optical signal is passed through a system of optical elements, e.g. collimator lenses, to amplify the signal and to align the impinging light for optimum detection [69, 70]. At the photodetector, i.e. a PD or an array of PDs, the optical signal is converted back to electrical current. The current

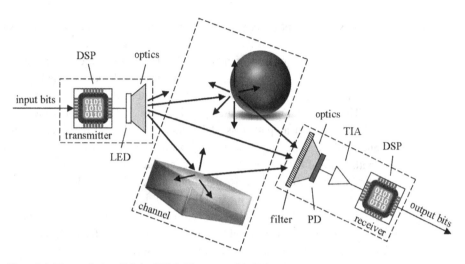

Figure 2.1 Transmission link in OWC. The general building blocks of the transmitter and receiver are presented. The optical wireless channel is illustrated, including several light rays and the reflecting objects in the setup

signal is electronically pre-amplified by means of a transimpedance amplifier (TIA). A DSP with an analog-to-digital converter (ADC) is employed for transformation of the analog current signal into a digital signal and demodulation of the information bits.

2.3   Communication scenarios

Originally, Kahn and Barry categorized the indoor OWC scenarios based on the relative strength between the LOS and NLOS signal components [21]. In addition, based on the relative directivity between the transmitter and the receiver, the links are classified as directed and non-directed. While directed LOS links provide high irradiation intensity at the receiver and a wide channel coherence bandwidth, they are not suitable for scenarios where high user mobility is required, and the link can be easily interrupted or blocked. On the other hand, even though non-directed NLOS communication provides a lower irradiation intensity at the receiver and a narrower channel coherence bandwidth, an NLOS scenario better supports user mobility, and it is significantly more robust to link blockage. Exploiting the reflection properties of the objects in the room, the irradiation intensity at the receiver in an NLOS communication setup can be enhanced by spot diffusing [71–73]. This is particularly beneficial for links in the infrared (IR) spectrum, where material reflectivity is generally higher compared to the visible light spectrum [16]. Since the primary functionality of visible light communication (VLC) links is illumination, a uniform light distribution over the topology of the room is desired [7]. Nonetheless, it is shown that the irradiation intensity at the receiver can be further enhanced by spot lighting [74]. There have been various studies on the optical wireless channel under different spatial conditions and optical configurations, based on

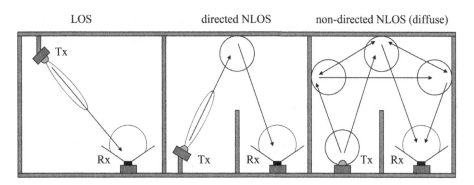

Figure 2.2 LOS and NLOS link configurations for OWC

direct measurements [70, 75–80] or ray-tracing simulations [21, 81–90]. It has been shown that the optical wireless channel can be well characterized by its RMS delay spread and path loss. RMS delay spread up to 13 ns and entire delay spread up to 100 ns have been reported for LOS and NLOS links, while optical path losses up to 80 dB can be experienced in an indoor setup.

2.3.1 Line-of-sight communication

The configuration of an LOS communication link is illustrated in Fig. 2.2. There is a direct path without obstruction and spatial alignment between the radiation pattern of the transmitter and the detection pattern of the receiver. In the following, an LOS component of the light propagation is referred to as the portion of the light radiated by the transmitter that arrives directly within the field of view (FOV) of the receiver.

2.3.2 Non-line-of-sight communication

Two common configurations of an NLOS communication link are discussed in the literature: directed NLOS and non-directed (diffuse) NLOS. These two setups are also illustrated in Fig. 2.2. Due to obstruction, e.g. a wall in the room, the transmitter and the receiver communicate by means of a single reflection on the ceiling or another wall in the room. In NLOS communication, the signal arrives at the receiver after one or multiple bounces off the objects in the room. The two NLOS scenarios are differentiated by the directivity of the transmitter. In the first case, directed NLOS, the transmitter has a very narrow radiation characteristic, projecting the light on a single spot on the ceiling which serves as a new transmitter. It relays the light to the receiver based on its reflection characteristic. This results in a single-reflection component with strong light intensity. The directed NLOS scenario is similar to spot diffusing [71–73]. In the second case, non-directed NLOS, the transmitter has a wide radiation characteristic, irradiating a large portion of the reflecting surface. In a closed room, it is likely that the radiated light will arrive at the receiver after one or multiple reflections on the surfaces, creating an ether for the light signal in the room. In the following, an NLOS component of the light propagation is referred to as the portion of the light radiated by the transmitter that arrives at the receiver after one or multiple reflections on the objects in the geometry of the room.

2.4   Optical front-ends

The optical front-ends of the transmitter and the receiver incorporate optical as well as electrical components, where the respective electrical-to-optical (E/O) and optical-to-electrical (O/E) conversions take place. In the following, the key components of the transmitter and the receiver are presented. Key parameters relevant for the system design are discussed, including common models for the radiation and detection characteristics, the optical spectral response of the transmitter and the receiver, the electrical modulation bandwidth, as well as the receiver noise.

2.4.1   Transmitter

Key characteristics of the transmitter include the radiation pattern, the optical spectral response, the E/O transfer characteristic, and the electrical modulation bandwidth. The radiation characteristic of a single LED is generally modeled by means of a generalized Lambertian radiation pattern [3, 21]. The FOV of an LED is defined as the angle between the points on the radiation pattern, where the directivity is reduced to 50%. In general, the FOV is specified by means of the semi-angle between the directions of maximum directivity and 50% directivity, $\theta_{FOV,Tx}$. It is often given in data sheets as $\pm\theta_{FOV,Tx}$. An incoherent diffuse LED can have an FOV in the range between $\pm10°$ and $\pm60°$, e.g. Vishay TSHG8200 or OSRAM LCW W5SM Golden Dragon [91, 92]. These two examples are given in Fig. 2.3. A transmitter can employ a single LED or an array of LEDs. Multiple LEDs can provide a higher radiant intensity in a given direction, when their radiation patterns are constructively aligned [93]. In addition, any radiation pattern can be obtained by co-locating LEDs in a non-planar fashion. Thus, omnidirectional non-planar transmitters can be built to serve as access points (APs) in a cellular network. A study on the deployment of a cellular OWC network inside an aircraft cabin is presented in Section 2.6. The design of the transmitter is subject to eye safety regulations. According to the BS EN 62471:2008 standard for photobiological safety of lamps and lamp systems [44], incoherent diffuse continuous-wave-modulated LEDs belong to the exempt group classification, and pose no photobiological hazard for the human eye if the irradiance does not exceed 100 W/m^2 at a distance of 0.2 m from the optical source in the direction of maximal directivity within 1000 s. However, as shown in Section 2.6, even very directive incoherent diffuse LEDs are inherently designed in accordance with the eye safety standard, and there is significant leeway for intensity amplification through an optical system of lenses and collimators.

In addition to the total output radiated optical power, the spectral response of the transmitter can be exploited in the design of high-capacity cellular OWC networks. For this purpose, the APs in the different cells can be tuned to different optical

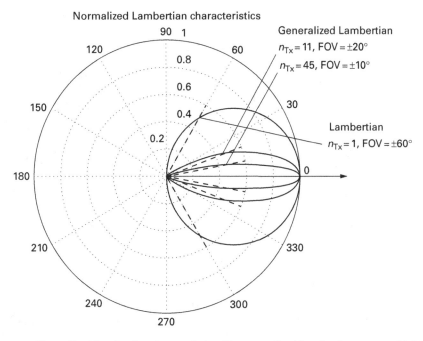

Figure 2.3 Normalized Lambertian characteristics. The normalized Lambertian pattern with FOV of $\pm 60°$ is illustrated, as well as two normalized generalized Lambertian patterns with FOVs of $\pm 10°$ and $\pm 20°$, including the respective Lambertian mode numbers, n_{Tx}

center wavelengths, providing non-interfering electrical modulation bandwidths. This approach is studied in detail in Chapter 7. In OWC, higher wavelength reuse factors can be employed without perceivable reduction in the total system capacity, since the electrical spectrum used by the OWC system, e.g. up to a few hundreds of MHz with state-of-the-art off-the-shelf components, is significantly smaller than the total electrical bandwidth available in the optical spectrum, e.g. approximately 670 THz. As a result, the electrical modulation bandwidth in the order of a few tens of MHz used at a given optical center frequency is significantly smaller than the distance between two optical center frequencies. Therefore, very high wavelength reuse factors can be accommodated with developments of optical and electrical filtering. Examples for optical spectral responses of IR, white, and colored LEDs are given in Fig. 2.4. A variety of incoherent diffuse IR LEDs can be found on the market at optical center frequencies of 830 nm, 850 nm, 870 nm, 890 nm, 940 nm, and 950 nm with a 3-dB bandwidth of approximately 40 nm [94]. White LEDs have a spectral emission in the range between 400 nm and 700 nm [92], while individual colors are obtainable with 3-dB wavelength bandwidths of approximately 20 nm. In general, white LEDs can be made as a mixture of red, green, and blue LEDs. More commonly though, white LEDs are made of blue LEDs with the addition of a yellow phosphor due to reduced fabrication complexity. However, because of the slow response of the yellow phosphor component, the modulation bandwidths of such LEDs are limited to a few MHz [25]. In order to overcome this limitation, a blue filter at the receiver is employed to filter out

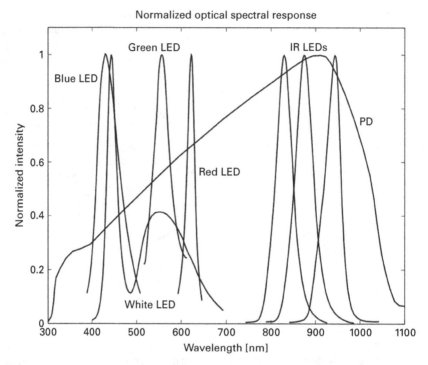

Figure 2.4 Optical spectral responses of LEDs and PDs in the visible and IR optical spectrum. Examples are given for white, colored, and IR LEDs, as well as a PD for both VLC and IR OWC

the slow yellow component. As a result, modulation bandwidths of several tens of MHz can be achieved [9, 11].

The non-linear transfer of input electrical power to output optical power is a further concern in the design of the transmitter, and it is analyzed in detail in Chapters 3 and 4. Depending on the optical center wavelength, the optical emitter can be modulated within a certain electrical bandwidth. Recently, 3-dB electrical modulation bandwidths up to 35 MHz have been reported for off-the-shelf inexpensive white LEDs [9, 11], while similar electrical bandwidths are typical for IR LEDs [6]. For the realization of gigabit/s OWC links, signals with a bandwidth much greater than the 3-dB modulation bandwidth of the LED are transmitted, such as 180 MHz in [11]. In order to achieve an acceptable bit-error rate (BER) below 10^{-3}, which is the limit for forward error correction (FEC) coding, channel equalization techniques are required. These include decision feedback equalization at the receiver in single-carrier transmission or bit and power loading at the transmitter with single-tap equalization at the receiver in multi-carrier transmission. The performance of the equalization techniques is presented in Section 2.5.4.

2.4.2   Receiver

Key characteristics of the receiver include the detection pattern, the optical spectral response, the O/E transfer characteristic, the electrical modulation bandwidth, and the noise figure. The detection characteristic of a single PD is generally modeled by means of a Lambertian detection pattern [3, 21], illustrated in Fig. 2.3. Similar to the LED, the FOV of a PD is defined as the angle between the points on the detection pattern, where the directivity is reduced to 50%. In addition, the FOV of the receiver can be further enhanced or constrained by a system of attached optical elements. For example, an optical concentrator in the form of collimator lenses can be employed to broaden the detection characteristic and enhance the strength of the received signal [70]. Therefore, the additional FOV parameter of the attached optical elements is commonly added to the detection pattern of the PD to model the overall detection characteristic of the receiver. The overall FOV semi-angle of the receiver is denoted as $\theta_{FOV,Rx}$. In addition, a large area PD, such as the Hamamatsu S6967 [95], can be used to collect more of the total radiated signal optical power at the expense of a reduced electrical modulation bandwidth due to the larger capacitance associated with the large-area element [17]. Alternatively, an array of PDs can be used to alleviate this trade-off. In addition to an increased photosensitive area without a reduction of the modulation bandwidth, multiple PDs can be co-located in a non-planar fashion, in order to obtain any detection pattern of interest. For example, omnidirectional non-planar receivers can be built into the APs of a cellular network such as the one studied in Section 2.6.

The optical spectral response of the receiver is determined by the spectral response of the PD and the spectral response of the optical filter. The spectral response of a PD is generally given by the responsivity parameter, i.e. the transfer factor of optical power to electrical current over the optical spectral range. Typical maximum responsivity values are within the range of $0.6 - 0.8$ A/W [95]. In addition, a PD is generally responsive over a wide optical spectral range which is larger than the spectral response of the LED. For example, spectral responses of PDs from 320 to 1100 nm can be found [95], spanning over the visible light and IR spectrum, as illustrated in Fig. 2.4. Therefore, optical filters are commonly employed to separate the individual optical channels, for example in the case of a cellular OWC network deployment. Optical filters with a narrow optical 3-dB bandwidth of ± 10 nm and a transmittance up to 0.8 can be found at a wide range of center wavelengths over the entire optical spectrum, e.g. Thorlabs bandpass filters [96].

At the receiver, there is a linear transfer between input optical power and output electrical power over a significantly larger dynamic range [95]. Signal clipping due to saturation of the PD can be avoided by means of the commonly employed automatic gain control (AGC). Furthermore, in a practical communication scenario it is unlikely that the signal is non-linearly distorted at the receiver because of the signal attenuation due to path loss. For instance, the linear range of a Vishay TEMD5110X01 PD reaches up to 0.2 mW of incident optical power at room temperature [97]. For a practical indoor path loss range between 50 dB and 80 dB [16, 66], the transmitter needs to radiate more than 20 W of optical power in order to drive the PD to saturation. Since such an amount significantly exceeds the limits imposed by the eye safety regulations [44], it is assumed that non-linear distortion occurs only at the transmitter.

The electrical modulation bandwidth of the receiver is determined by the electrical bandwidths of the PD and the TIA. The electrical bandwidth of the receiver is generally larger than the electrical bandwidth of the transmitter. Electrical bandwidths in excess of 100 MHz have been reported [11, 98].

In OWC systems, the ambient light produces high-intensity shot noise at the receiver. In addition, thermal noise arises due to the electronic pre-amplifier in the receiver front-end, i.e. the TIA [21]. These two noise components dominate the additive noise at the receiver, w, which can be modeled as a random process with a zero-mean real-valued Gaussian distribution [21]. At the received constellation, the noise is perceived as a complex-valued additive white Gaussian noise (AWGN) with a double-sided electrical power spectral density (PSD) of $N_0/2$ per signal dimension [99]. Therefore, in M-PPM and M-PAM, w has an electrical PSD of $N_0/2$ and an electrical power of $\sigma^2_{\mathrm{AWGN}} = BN_0/2$. Here, B is the 3-dB double-sided electrical signal bandwidth of interest. In O-OFDM with M-QAM, the electrical PSD of w amounts to N_0 for an electrical power of $\sigma^2_{\mathrm{AWGN}} = BN_0$ because of the two-dimensional constellation [99]. For a given front-end setup, the electrical power of the AWGN can be expressed as a function of the wavelength, λ, as follows:

$$\sigma^2_{\mathrm{AWGN}}(\lambda) = 2q(P_{\mathrm{R}}(\lambda) + P_{\mathrm{bg}}(\lambda))S_{\mathrm{PD}}(\lambda)G_{\mathrm{TIA}}T_{\mathrm{OF}}(\lambda)G_{\mathrm{OC}}B + 4k_{\mathrm{B}}TB . \quad (2.1)$$

The first addend represents the shot noise with a contribution from the optical signal intensity and the background illumination, while the second addend is the thermal noise component. Here, $q = 1.6 \times 10^{-19}$ C is the elementary electric charge, $P_{\mathrm{R}}(\lambda)$ is the optical signal power impinging on the receiver, and $P_{\mathrm{bg}}(\lambda)$ is the optical power of the background illumination, i.e. the ambient light in the room. The responsivity of the PD is denoted by $S_{\mathrm{PD}}(\lambda)$, G_{TIA} is the gain of the TIA, $T_{\mathrm{OF}}(\lambda)$ is the transmittance of the optical filter, G_{OC} is the gain of the optical concentrator, k_{B} is the Boltzmann's constant, and T is the absolute temperature.

There are several techniques that contribute to the increase of the electrical SNR at the receiver. In general, because of the fact that the AWGN is directly proportional to the modulation bandwidth, a receiver matched to the transmitted signal in terms of modulation bandwidth can contribute to the reduction of AWGN, and, therefore, to increased SNR and throughput of the OWC system. Alternatively, a matched electrical filter is employed. In addition, an optical filter can be employed to remove the background light component, and, therefore, to reduce the shot noise. When the ambient-induced shot noise is reduced, the deployment of an avalanche PD can provide a significant enhancement of the SNR as compared to an ordinary positive–intrinsic–negative (PIN) PD due to its internal electrical gain at high reverse bias [21]. Lastly, a large gain TIA, e.g. Analog Devices AD8015 TIA [100], with a low noise figure can help maintain a low AWGN power, while providing a significant signal amplification.

2.5 Optical wireless channel

The optical wireless channel has been shown to be a linear, time-invariant, memoryless system with an impulse response of a finite duration [21]. Intensity modulation and direct detection (IM/DD) is the most common technique for OWC. Here, the amplitude

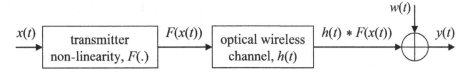

Figure 2.5 Generalized block diagram of the OWC link in the time domain. It includes the LED non-linearity, the dispersive optical wireless channel, and the AWGN

of the received current signal is proportional to the integral of the incident optical power over the area of the PD. In general, since the area of a PD is millions of square wavelengths, there is a rich spatial diversity. As a result, there is no fast fading in OWC, but only slow fading in the form of shadowing. In the following, key models for OWC are presented, including the model for the communication link in the presence of receiver noise, models for the channel path loss and delay spread, as well as techniques for channel equalization at the receiver.

2.5.1 Channel model

OWC can be described by the following continuous-time model for a noisy communication link:

$$y = h * F(x) + w ,$$ (2.2)

where $y(t)$ represents the received distorted replica of the transmitted signal, $x(t)$, which is subjected to the non-linear distortion function, $F(x(t))$, of the transmitter front-end. The non-linearly distorted transmitted signal is convolved with the channel impulse response, $h(t)$, and it is distorted by AWGN, $w(t)$, at the receiver. Here, $*$ denotes linear convolution. The generalized model of the OWC link in the time domain is illustrated in Fig. 2.5. Since the OWC system is implemented by means of a DSP, the following equivalent discrete model for a noisy communication link is employed in the system description:

$$\mathbf{y} = \mathbf{h} \star F(\mathbf{x}) + \mathbf{w} ,$$ (2.3)

where $\star$ denotes discrete linear convolution. Here, the transmitted signal vector, $\mathbf{x}$, contains $Z_\mathbf{x}$ samples, the channel impulse response vector, $\mathbf{h}$, has $Z_\mathbf{h}$ samples, and, as a result, the AWGN vector, $\mathbf{w}$, and the received signal vector, $\mathbf{y}$, have $Z_\mathbf{x} + Z_\mathbf{h} - 1$ samples [99]. The discrete signal vectors are obtained by sampling of the equivalent continuous-time signals. The sampling rates over a time period of T differ in the considered systems, and the details are presented in Section 3.2.

2.5.2 Path loss

The modeling of the path loss of the optical wireless channel gained significant interest after the pioneering work of Gfeller and Bapst [3]. They presented an analytical model for the received optical power in LOS and single-reflection NLOS OWC. An illustration of the geometry of this communication scenario is given in Fig. 2.6. Parameters

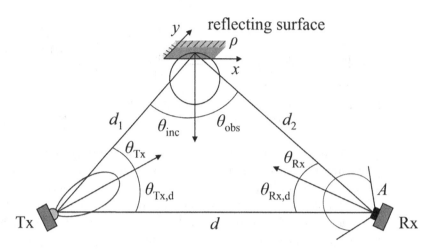

Figure 2.6 Geometry of an OWC scenario, including an LOS and a single-reflection NLOS component. The mutual orientation of the transmitter and the receiver, as well as their orientation towards the reflecting surface, are described by means of observation and incident angles with respect to the normal direction

describing the mutual orientation of the transmitter and the receiver, as well as their orientation towards the reflecting surface, are included. As a common trend in literature, the transmitter is modeled by means of a generalized Lambertian radiation pattern, while the receiver has a Lambertian detection pattern with a given FOV. As a result, the optical power impinging on the PD, P_R, can be generalized in the case of LOS and single-reflection propagation for any given bi-directional reflectance distribution function (BRDF), $\mathcal{R}(\theta, \phi)$, as follows:

$$
P_R = P_T \frac{n_{Tx} + 1}{2\pi} \cos^{n_{Tx}}(\theta_{Tx,d}) \frac{A}{d^2} \cos(\theta_{Rx,d}) \text{rect}(\theta_{Rx,d})
$$

$$
+ P_T \frac{n_{Tx} + 1}{2\pi} \int_x \int_y \int_\theta \int_\phi \cos^{n_{Tx}}(\theta_{Tx}) \frac{\rho \mathcal{R}(\theta, \phi)}{d_1^2} \tag{2.4}
$$

$$
\times \frac{A}{d_2^2} \cos(\theta_{Rx}) \text{rect}(\theta_{Rx}) \, dx \, dy \, d\theta \, d\phi \, .
$$

Here, P_T is the total radiated optical power of the LED, and n_{Tx} denotes the Lambertian mode number of the radiation lobe related to the LED directivity. The first addend represents the LOS received optical power from the direct path, while the second addend calculates the NLOS received optical power after a single reflection on the reflecting surface, taking into account the BRDF. Here, the received optical power is obtained after integration of the reflective surface in x and y directions and integration of the BRDF over θ and ϕ angles, ranging over $0 \leq \theta \leq \pi/2$ and $0 \leq \phi \leq 2\pi$. In addition, $\theta_{Tx,d}$ denotes the observation angle of the transmitter on the direct path, and $\theta_{Rx,d}$ is the direct incident angle of the receiver. Observation and incident angles are computed with respect to the normal directions of the radiating, reflecting, or detecting elements. The distance between transmitter and receiver on the direct path is given by d, and A is the photosensitive area of the PD. On the non-direct path, θ_{Tx} is the observation angle of the transmitter towards the reflective surface, θ_{Rx} is the incident angle of the receiver from the reflective surface, and ρ denotes the reflection coefficient of the surface. The distance between the transmitter and the reflective surface is given by d_1, while d_2 is the distance between the receiver and the reflective surface. Lastly, $\text{rect}(\theta)$ is related to the receiver FOV as follows:

$$\text{rect}(\theta) = \begin{cases} 1, & \text{if } \theta \leq \theta_{\text{FOV,Rx}}, \\ 0, & \text{if } \theta > \theta_{\text{FOV,Rx}}. \end{cases} \tag{2.5}$$

In literature, the BRDF, $\mathcal{R}(\theta, \phi)$, is often abstracted only as a function of θ for simplicity. Gfeller and Bapst considered a Lambertian pattern for the spatial attenuation distribution of the incident and reflected directions [3]:

$$\mathcal{R}(\theta_{\text{inc}}, \theta_{\text{obs}}) = \cos(\theta_{\text{inc}}) \frac{\cos(\theta_{\text{obs}})}{\pi} , \tag{2.6}$$

where θ_{inc} denotes the incident angle of the incoming light, and θ_{obs} is the observation angle of the outgoing light.

Later on, the reflection model was further abstracted to only consider a Lambertian pattern for the spatial distribution of the outgoing light, while any incident angle of the incoming light was permitted [21]:

$$\mathcal{R}(\theta_{\text{inc}}) = \frac{\cos(\theta_{\text{obs}})}{\pi}. \tag{2.7}$$

Using this model, analytical models for directed and non-directed (diffuse) NLOS communication have been derived, considering a single reflection of a wall or ceiling [21]. Alternatively, the Phong reflection model can be used, taking into account also a specular reflection [86, 101]:

$$\mathcal{R}(\theta_{\text{inc}}, \theta_{\text{obs}}) = \frac{\eta}{\pi} \cos(\theta_{\text{obs}}) + (1 - \eta) \frac{n_{\text{spec}} + 1}{2\pi} \cos^{n_{\text{spec}}}(\theta_{\text{obs}} - \theta_{\text{inc}}) . \tag{2.8}$$

Here, the specular component of the light outgoing from the reflecting surface is modeled according to a generalized Lambertian pattern. The mode number of the specular reflection lobe is denoted by n_{spec}, and η is the fraction of the light reflected in a diffuse Lambertian fashion.

This continuous model for the received optical power in an OWC link is applicable when the spectral response of the receiver is much larger than the spectral response of the transmitter, i.e. it can be considered as a constant. This is generally the case for IR LEDs and PDs, where the IR emittance of the LED can be assumed to be concentrated around a single wavelength and the PD covers a much wider spectral range. As a result, there is no dependance on the wavelength. However, in VLC, a white LED has a broad spectrum of emittance, and therefore the spectral response of the receiver is comparable to the spectral response of the transmitter. As a result, (2.4) is a function of the wavelength, λ. In addition, the geometry of an indoor VLC setup, such as an aircraft cabin [16], has complex reflecting surfaces on the various objects, such as seats, baggage compartments, and sidewall panels. Therefore, light propagation can undergo multiple reflections, and ray-tracing is commonly employed [16, 102].

Ray-tracing was introduced as a deterministic algorithm in [103]. It was later recognized as a numerical solution to the rendering equation [104], and the Monte Carlo ray-tracing algorithm was developed for image-based rendering, i.e. backward ray-tracing from the viewpoint with FOV towards the light source in the geometry or a partial solution to the global illumination problem. The object-based or the forward MCRT from the light source towards the viewpoint was introduced in [105] as the solution to the global illumination problem. The deterministic ray-tracing and the MCRT algorithms have been applied to OWC, where the received power is only stored for the PD as viewpoint and not for the surfaces in the geometry [21, 83].

In the deterministic algorithm, the radiation characteristic of the LED is discretized in a number of rays with power of P_T. The rays are traced sequentially, while each ray performs a random walk within the geometry. In the MCRT algorithm, a number of rays, N_{rays}, with power of P_T/N_{rays}, are randomly generated according to a spatial distribution following the radiation characteristic of the LED. The rays are traced in a random sequence. The LOS rays arrive directly at the PD and their contribution to the received power is given by the LOS component in (2.13). The NLOS rays undergo reflections before arriving at the PD. When a ray hits a reflecting surface, the power of the rays is decreased by the reflection coefficient and the incoming BRDF of the surface. The surface is transformed into a light source with power equal to the attenuated power of the ray, and the direct contribution of this new light source to the PD is calculated. Then, the rays are extinguished, and a new outgoing ray with an equal attenuated power is generated according to a spatial distribution following the BRDF of the surface. The MCRT continues until a criterion for stochastic accuracy, or relative error, is met. For the calculation of this criterion, the rays are randomly allocated to bins, and the error is represented as the percentage of the variance of the bins to the mean-squared value [102]. In general, the MCRT algorithm achieves a given stochastic accuracy with significantly lower computational effort.

The received power according to the deterministic ray-tracing algorithm can be obtained from (2.4), which is generalized to accommodate multiple reflections, and integration is discretized as follows:

$$P_R(\lambda) = \sum_{i=1}^{N_{LOS}} \frac{P_{LOS,i}(\lambda)}{N_{LOS}} + \sum_{j=1}^{N_{NLOS}} \frac{P_{NLOS,j}(\lambda)}{N_{NLOS}} , \tag{2.9}$$

where

$$P_{LOS,i}(\lambda) = P_T(\lambda) \frac{n_{Tx}+1}{2\pi} \cos^{n_{Tx}}(\theta_{Tx,i}) \frac{A}{d_i^2} \cos(\theta_{Rx,i}) \mathrm{rect}(\theta_{Rx,i}) \tag{2.10}$$

and

$$P_{NLOS,j}(\lambda) = P_T(\lambda) \frac{n_{Tx}+1}{2\pi} \cos^{n_{Tx}}(\theta_{Tx,j}) \left(\prod_{k=1}^{N_{refl,j}} \rho_k(\lambda) \frac{\mathcal{R}_k(\theta_k, \phi_k, \lambda)}{d_k^2} \right) \tag{2.11}$$

$$\times \frac{A}{d_j^2} \cos(\theta_{Rx,j}) \mathrm{rect}(\theta_{Rx,j}) .$$

Here, N_{LOS} is the number of LOS rays directly impinging on the PD, and N_{LOS} is the number of rays which undergo one or multiple reflections before reaching the PD. In the LOS component, d_i is the distance the ith ray travels to the PD, $\theta_{Tx,i}$ is the observation angle of the transmitter towards the receiver on the ith ray, while $\theta_{Rx,i}$ is the associated incident angle of the receiver. In the NLOS component, $\theta_{Tx,j}$ is the observation angle of the transmitter towards the first reflection on the jth ray, and $\theta_{Rx,j}$ is the incident angle of the receiver from the last reflection. The number of reflections that the jth ray undergoes is denoted by $N_{refl,j}$. A measured BRDF of the reflecting surface can be used [102]. Otherwise, the simplified reflection models from (2.6), (2.7), and (2.8) can be employed, where the reflection parameters, $\rho_k(\lambda)$, $\eta_k(\lambda)$, and $n_{spec,k}(\lambda)$ are specified for every surface and every wavelength. The distance between the previous reflection of the jth ray, or the LED in the case of a single reflection, to the kth reflection is denoted by d_k, and d_j is the distance from the last reflection to the receiver. Furthermore, when $\sqrt{A} << d$, the LOS component can be approximated as follows:

$$\sum_{i=1}^{N_{LOS}} \frac{P_{LOS,i}(\lambda)}{N_{LOS}} = P_T(\lambda)\frac{n_{Tx}+1}{2\pi}\cos^{n_{Tx}}(\theta_{Tx,d})\frac{A}{d^2}\cos(\theta_{Rx,d})\text{rect}(\theta_{Rx,d}). \tag{2.12}$$

The received power according to the MCRT algorithm can be expressed by modifying (2.9) as follows:

$$P_R(\lambda) = P_T(\lambda)\frac{n_{Tx}+1}{2\pi}\cos^{n_{Tx}}(\theta_{Tx,d})\frac{A}{d^2}\cos(\theta_{Rx,d})\text{rect}(\theta_{Rx,d})$$

$$+ \frac{P_T(\lambda)}{N_{rays}}\sum_{i=1}^{N_{NLOS}}\sum_{j=1}^{N_{refl}}\left(\prod_{k=1}^{j-1}\rho_k(\lambda)\mathcal{R}_{in,k}(\theta_k,\phi_k,\lambda)\right)\rho_j(\lambda)\mathcal{R}_j(\theta_j,\phi_j,\lambda) \tag{2.13}$$

$$\times \frac{A}{d_j^2}\cos(\theta_{Rx,j})\text{rect}(\theta_{Rx,j}) .$$

Here, the first addend is the LOS component, while the second addend represents the NLOS component. The total number of the traced rays is given by $N_{rays} = N_{LOS} + N_{NLOS}$, and N_{refl} denotes the total number of reflecting surfaces. The portion of the BRDF related to the incoming light at the jth reflection is denoted by $\mathcal{R}_{in,j}(\theta_j,\phi_j,\lambda)$. It can be readily obtained from the measured BRDF. If the simplified reflection models are used, $\mathcal{R}_{in,j}(\theta_j,\phi_j,\lambda)$ is a Lambertian function according to (2.6) or a unity factor according to (2.7) and (2.8).

While performing ray-tracing, the pair (P_R, d_{tot}) can be computed for each ray, where d_{tot} is the total distance (including the reflections) a ray travels to the PD. As a result, the arrival time of the ray can be calculated as follows: d_{tot}/c, where c is the speed of light. The received power of the ray can be accumulated in bins according to the arrival time in order to obtain the impulse response of the channel [21].

Of major importance for the simulation of different OWC system configurations is the establishment of a generic statistical path loss model. The measurements conducted in [78] suggest that variations of channel path loss are smooth, and a simple curve-fitting algorithm can be used to accurately approximate the intermediate values. Therefore, the already existing model for path loss can be adopted from the radio frequency (RF) transmission. Thus, the path loss at distance d, $\mathrm{PL}(d)$, can be presented in the logarithmic domain as follows [16, 66, 106]:

$$\mathrm{PL}(d) = 10 \log_{10}\left(\frac{P_T}{P_R(d)}\right) = \mathrm{PL}(d_{\mathrm{ref}}) + 10\zeta \log_{10}\left(\frac{d}{d_{\mathrm{ref}}}\right) + \xi , \qquad (2.14)$$

where $\mathrm{PL}(d_{\mathrm{ref}})$ is the path loss at reference distance d_{ref}, ζ is the path loss exponent, and ξ is a random variable that accounts for shadowing effects. In the logarithmic domain, ξ has a zero-mean Gaussian distribution which corresponds to a log-normal distribution in the linear domain. The main task is the determination of the path loss exponent and the characterization of the shadowing component for the particular system. The validity of this model is to be demonstrated in an experimental setup in Section 2.6, where an MCRT simulation is employed, in order to assess the path loss statistics. However, since the chosen simulation software provides an output for irradiance at a chosen location, the path loss equation model is adjusted accordingly to incorporate the receiver model with a variable photosensitive area. The modified path loss model is given as follows:

$$\mathrm{PL}(d) = 10 \log_{10}\left(\frac{P_T}{E(d_{\mathrm{ref}})A_0}\right) - 10 \log_{10}\left(\frac{A}{A_0}\right) + 10\zeta \log_{10}\left(\frac{d}{d_{\mathrm{ref}}}\right) + \xi . \qquad (2.15)$$

The irradiance at the reference distance d_{ref} is given by $E(d_{\mathrm{ref}})$ and A_0 is 1 m^2. In analogy to the aperture of an RF antenna, the photosensitive area A of the PD is a scaling factor of the received optical intensity.

The optical path gain parameter to be used for the modeling of the communication link in (2.2) and (2.3) can be obtained from the inverse of the optical path loss, including some additional parameters for the O/E conversion. Therefore, the overall optical path gain per optical wavelength can be expressed as follows:

$$g_{h(\mathrm{opt})}(\lambda) = P_R(\lambda)S_{\mathrm{PD}}(\lambda)G_{\mathrm{TIA}}T_{\mathrm{OF}}(\lambda)G_{\mathrm{OC}}/(P_T(\lambda)\sqrt{R_{\mathrm{load}}}) , \qquad (2.16)$$

where R_{load} is the load resistance over which the received signal is measured. In general, for a given OWC setup with a considered optical bandwidth, the average optical path gain is obtained by averaging the optical path gains per wavelength over the optical bandwidth. In addition, the optical path gain can be related to the electrical path gain, $g_{h(\mathrm{elec})}$, as follows:

$$g_{h(\mathrm{elec})} = \frac{1}{B}\int_{-B/2}^{B/2}|H(f)|^2\,df = g_{h(\mathrm{opt})}^2\frac{1}{B}\int_{-B/2}^{B/2}|H_{\mathrm{norm}}(f)|^2\,df . \qquad (2.17)$$

Here, $H(f)$ is the Fourier transform of the impulse response of the optical wireless channel, $h(t)$, $H_{\mathrm{norm}}(f)$ is the Fourier transform of the normalized impulse response, $h_{\mathrm{norm}}(t)$.

2.5.3   Delay spread and coherence bandwidth

There is an inverse relation between the channel delay spread and the channel bandwidth. Also, it is shown in [84] that the variation of the received power in environments with different reflectivity and the available bandwidth are almost inverse to each

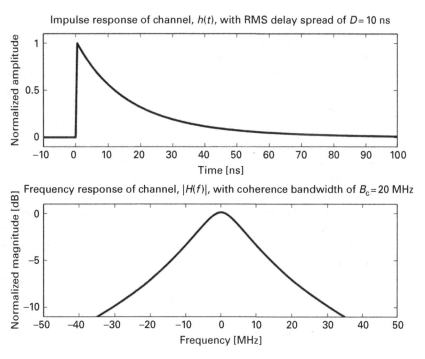

Figure 2.7 An example of the model for the impulse response and the corresponding frequency response of the optical wireless channel

other. This means that in a highly reflective geometry, low path loss is expected, which increases the power of the rays coming from different propagation paths, resulting in high delay spread and low channel bandwidth. In contrast, if the geometry reflectivity is low, high path loss is expected which leads to low delay spread and high channel bandwidth.

LOS and NLOS optical wireless channels can be accurately modeled by the rapidly decaying exponential impulse response function as follows [21]:

$$h(t) = g_{h(opt)} h_{norm}(t) = g_{h(opt)} U(t) \frac{6a^6}{(t+a)^7} . \tag{2.18}$$

Here, $U(t)$ stands for the unit step function, and a is related to the RMS delay spread, D, as follows: $a = D\sqrt{11/13}$. The 3-dB coherence bandwidth of the channel can be expressed from the RMS delay spread as follows: $B_c = 1/(5D)$ [106]. RMS delay spreads between 1.3 ns and 12 ns are reported for LOS links, whereas RMS delay spreads between 7 ns and 13 ns are reported for NLOS links [21]. A realization of the channel impulse response with an RMS delay spread of 10 ns and the corresponding frequency response with a coherence bandwidth of 20 MHz are presented in Fig. 2.7.

2.5.4 Channel equalization

In single-carrier OWC systems, such as M-PPM and M-PAM, the RMS delay spread of the channel becomes comparable to or larger than the pulse duration at high data rates which causes severe inter-symbol interference (ISI). Equivalently, the signal bandwidth exceeds the channel coherence bandwidth. In general, similar effects are also caused by the low-pass frequency response of the front-end components, such as LEDs, PDs, and amplifiers. As a result, the BER performance is degraded, and the systems effectively incur an SNR penalty. In practical system implementations, a multi-tap linear FFE or a non-linear DFE with ZF or MMSE criteria are deployed to reduce the SNR penalty. The theoretical limits for the least SNR penalty of the equalizers are discussed below. Because of the fact that in single-carrier modulation, the RMS delay spread of the channel is comparable to or larger than the pulse duration at high data rates, a large number of channel taps are required to accurately represent the impulse response function in order to counter its effect. As a result, the computational complexity of the equalization process is significantly increased. In addition, since the number of channel taps depends on the frequency at which the received signal, y, is sampled, the received pulses are often oversampled in order to increase the accuracy of the equalization process, leading to even higher computational effort. However, these approaches are fundamentally limited by the maximum sampling frequency of the ADC employed. As a result, the theoretical limits for the SNR penalty are not generally achievable in practical system implementations. The gain of the equalizer is represented by the gain factor, G_{EQ}. For a linear FFE with the ZF criterion, the gain factor is given as follows [45]:

$$G_{EQ} = \left(\frac{1}{B} \int_{-B/2}^{B/2} \frac{1}{\Lambda(f)} \, df \right)^{-1}. \tag{2.19}$$

Here, $\Lambda(f)$ is expressed as follows [45]:

$$\Lambda(f) = \sum_{z=-\infty}^{\infty} |V(f - zB)H(f - zB)|^2, \tag{2.20}$$

where $V(f)$ is the Fourier transform of the impulse response of the pulse shaping filter at the transmitter, $v(t)$, and z is the counter in the summation. For a linear FFE with the MMSE criterion, the gain factor is given as follows [45]:

$$G_{EQ} = \left(\frac{1}{B} \int_{-B/2}^{B/2} \frac{1}{\Lambda(f) + \gamma_{b(elec)}^{-1}} \, df \right)^{-1}. \tag{2.21}$$

For a non-linear DFE with the ZF criterion, the gain factor is given as follows [45]:

$$G_{EQ} = \exp \left(\frac{1}{B} \int_{-B/2}^{B/2} \ln \left(\Lambda(f) \right) df \right). \tag{2.22}$$

For a non-linear DFE with the MMSE criterion, the gain factor is given as follows [45]:

$$G_{EQ} = \exp \left(\frac{1}{B} \int_{-B/2}^{B/2} \ln \left(\Lambda(f) + \gamma_{b(elec)}^{-1} \right) df \right). \tag{2.23}$$

The gain factor G_{EQ} represents the theoretical lower bound for the electrical SNR penalty, which the BER performance incurs at high data rates. This lower bound is achieved when an infinite number of channel taps are considered in the FFE and DFE, which is not generally achievable in practice. In general, ZF eliminates the ISI completely, but it results in AWGN amplification when the path gain decreases. Alternatively, MMSE equalization alleviates this issue as it provides a trade-off between AWGN amplification and residual ISI. Since, however, an equalizer with an MMSE criterion requires higher computational effort, and it only reduces the SNR penalty by approximately 0.5 dB as compared to the ZF criterion for the considered channel model, ZF is considered throughout this book.

In multi-carrier systems such as OFDM-based OWC, the RMS delay spread is significantly shorter than the symbol duration, and therefore the equalization process is considerably simplified to single-tap equalization [107]. The ISI and the inter-carrier interference (ICI) are completely eliminated by the use of a large number of subcarriers and a cyclic prefix (CP) of length which is at least equal to, or larger than, the maximum excess delay of the channel. Thus, the dispersive optical wireless channel is transformed into a flat fading channel over the subcarrier bandwidth [107]. The CP is an additional overhead, which generally results in some loss of spectral efficiency. However, since the delay spread in OWC is much less than in RF systems, the expected spectral efficiency loss is insignificant [108], in particular if the CP, for example, is exploited for channel estimation [109]. For example, the ISI from maximum delay spreads of up to 100 ns can be compensated by a CP of 2 samples at a sampling rate of 20 MHz. For a fast Fourier transform (FFT) size of 1024, this results in a negligible increase of the electrical SNR requirement with 0.01 dB and a reduction of the spectral efficiency of 0.2%. Therefore, the channel can be safely considered as flat fading over the entire OFDM frame for bandwidths up to 20 MHz [21, 107], and it can be primarily characterized by the optical path gain coefficient, $g_{h(opt)}$ [16, 66]. A large number of subcarriers, e.g. greater than 64, also ensures that the time-domain signal follows a close to Gaussian distribution [110]. This assumption greatly simplifies the derivation of the OFDM signal statistics. In addition, the CP transforms the linear convolution with the channel into a cyclic convolution, facilitating a single-tap linear FFE and eliminating the need for a non-linear DFE. Even though the channel can be considered as flat fading over the individual subcarriers, the non-flat channel frequency response over the entire OFDM frame in the case of a larger frame bandwidth still leads to an SNR penalty for the average frame BER. Here, the single-tap equalizer is generally paired with bit and power loading [49, 50], in order to minimize this SNR penalty. The gain factor of the equalizer, G_{EQ}, can be obtained by means of a Monte Carlo simulation. The time dispersion of the channel is, therefore, a less critical parameter when using OFDM as opposed to when using pulsed modulation schemes, such as M-PPM and M-PAM.

2.6 Cellular network: a case study in an aircraft cabin

OWC systems are a preferred solution in areas where the interference with RF-based technology is an issue such as in aircraft cabins or hospitals. The use of this technology within aircraft cabins has additional advantages. OWC can support a variety of applications, ranging from onboard inter-system communication, flight maintenance on the ground, as well as flight entertainment. Significant amounts of cabling can be saved, which results in reduced weight and greater flexibility in cabin layout designs. In this section, the deployment of an optical wireless network within an aircraft cabin is considered [16–18, 33, 66, 111]. The case of aircraft use is an example to help understand the basic limitations in an indoor environment which is characterized by a large number of users within a very limited space – an ideal scenario for OWC. Several other studies have investigated optical wireless networks for aircraft cabins with the purpose of intra-cabin passenger communication [112–114]. Measurement results reported in these studies demonstrate that data rates of several tens of Mbps are feasible within the cabin. Results on achievable channel bandwidth inside an aircraft cabin are reported in [70]. It is shown that a channel bandwidth of more than 50 MHz is possible through optimization of irradiation and receiver FOV.

In order to increase the capacity of a wireless system, a cellular network can be deployed [106]. Comprehensive spatial SIR maps over the cabin topology are demonstrated in [16] by means of a Monte Carlo ray-tracing irradiation simulation for different wavelength reuse scenarios. By the use of a multi-beam transmitter [6, 115, 116], any radiation pattern can be obtained by combining Lambertian LEDs. In addition, light shaping diffusors can provide an even cell coverage [117] and reduce the RMS delay spread [118]. Optimization of the radiation/detection patterns to increase the irradiance in a given direction is discussed in [111]. These techniques enable the maximization of cell coverage in the cellular OWC network.

In this section, first, path loss models are developed for optical wireless transmission inside an aircraft cabin. Second, SIR maps in a cellular network are determined via simulation. For this purpose, an MCRT simulation is performed in a geometric computer-aided design (CAD) cabin model with defined position, azimuth (AZ), elevation (EL), and FOV properties of transmitters and receivers. Mathematical models are developed for LOS and NLOS path losses along particular paths, including estimation of the path loss exponent and the shadowing component. The shadowing is modeled as a random variable from a Gaussian distribution with a zero-mean and standard deviation σ_{shad} in the logarithmic domain. In the linear domain, the shadowing follows a log-normal distribution. The validity of this model is confirmed. The two LOS scenarios show that the irradiance distribution under LOS conditions experiences an attenuation with a path loss exponent of 1.92 and 1.94 and a standard deviation of the shadowing component of 0.81 dB and 0.57 dB, respectively. This results in an LOS path loss between 27 dB and 69 dB over a distance of between 0.04 m and 5.6 m. In NLOS conditions, however, the path loss exponent varies, depending on the nature of the NLOS cases considered. The presented NLOS scenarios yield path loss exponent values of 2.26 and 1.28, and shadowing standard deviation values of 1.27 dB and 0.7 dB, respectively. This results in an NLOS path loss between 54 dB and 80 dB over a distance of between 0.39 m and 5.61 m in the first case, while in the second case the NLOS path loss is between 64 dB and 72 dB over a distance of between 1.45 m and 5.78 m. Since the latter NLOS path loss is comparable to the LOS path loss at larger distances, an NLOS setup with higher robustness to link blockage and a better mobility service is a highly relevant communication scenario for OWC. Finally, the cabin is divided into cells and SIR maps are presented for different wavelength reuse factors. It is shown that at the edges of the circular cells with a diameter of 2.8 m, an SIR of -5.5 dB is achieved in a horizontal cross-section of the cabin for wavelength reuse of 1, and -2 dB and 3 dB for wavelength reuse factors of 2 and 3, respectively. This means that in an aircraft cabin, for reuse factors less than three, viable communication at the cell edges is not feasible without additional interference avoidance or interference mitigation techniques. Alternatively, narrow-band filters and collimation lenses can be used at the receiver side to enhance the SIR.

2.6.1 Ray-tracing for signal and interference modeling

Modeling the propagation of an optical wireless signal in an indoor environment is a challenging task. The reflection characteristics of the surfaces in the setup highly influence the signal power distribution. Moreover, any change in position, AZ, EL, or FOV of a transmitter and/or receiver affects the channel characteristics. In addition, movement, blockage, and shadowing also vary the properties of the channel.

A key contribution is the determination of the SIR distribution of an optical cellular network inside the cabin of an aircraft. A study conducted in [63] shows that a configuration of hexagonal cells and channel reuse of 3 is co-channel-interference-(CCI)-limited for small cell radii and noise-limited for large cell radii, but it does not provide a comprehensive spatial SIR analysis. Indoor peer-to-peer and client/server wireless topologies have been analyzed in [68], but a noise-limited network is assumed in this work. However, due to the expected high link density in an aircraft cabin, the wireless network is interference-limited [112, 113]. Therefore, in this section, SIR maps for different wavelength reuse factors are determined in order to quantify the minimal achievable SIR at the cell edges, which generally experience the highest CCI.

In order to determine the SIR maps, it is important to establish a model for the path loss in such an environment. The path loss in an optical wireless communication system has been studied in [68, 70, 75, 77, 78, 85, 90]. The major challenge in path loss modeling is related to the complexity of a global illumination simulation, involving a geometric model of a setup, materials reflection properties, definition of the radiation pattern of the transmitters, and the detection pattern of the family of receivers. Measurements and simulations conducted in [68, 77, 85] show that optical path loss ranges between 50 dB and 80 dB for single-reflection, unshadowed, or shadowed diffuse channels. Further measurements in [78] suggest that the path loss variation along the path is smooth, and simple curve-fitting can be used to accurately estimate the intermediate values. A radiosity simulation method described in [90] can be used to determine the path loss for a particular transmitter and receiver placement in a room.

However, what the current approaches generally lack is an automated way to determine the path loss in a chosen room or setup comprehensively at every possible position of interest. There is a need for a statistical model of the optical wireless path loss as a function of distance in different indoor setups, such as the cabin of an aircraft. Such a flexible model would enable the optimization of the position and the parameters of the transmitters and receivers, such as AZ, EL, and FOV. In addition, SIR maps for different Tx and Rx configurations and different wavelength reuse factors for such an environment are still an open issue.

In this book, an MCRT approach is used to obtain the optical wireless path loss distribution within an aircraft cabin. A global irradiation simulation is performed with the software tool Specter by Integra Inc [102]. This software utilizes an MCRT algorithm as a primary lighting simulation method. The following inputs are required for an MCRT irradiation simulation: geometric models of the objects in the setup, definition of the reflection characteristics of the materials on these objects, and definition of key properties of light sources and observers. A particular potential deployment scenario is considered, based upon a cellular optical wireless network. This is realized through a combination of transmitter and receiver positions, AZ, EL, and FOV of selected off-the-shelf components. From the spatial irradiation distribution in the aircraft cabin, mathematical models for the path loss in LOS and NLOS communication scenarios are obtained. It is confirmed that the optical path loss is linear over logarithmic distance. This allows for the establishment of a statistical model of the path loss as a function of distance along chosen paths for the predefined light sources. Based on the MCRT path loss distribution calculations, the SIR distribution in the setup is determined, assuming different wavelength reuse factors. Finally, the SIR maps in the deployment scenario are verified with the mathematical path loss model.

2.6.2 Cabin setup: propagation paths, cellular configuration, and wavelength reuse

The conducted study aims at estimating the optical wireless channel path loss for LOS and NLOS communication scenarios in a cellular network inside an aircraft cabin and the computation of SIR maps over the topology. Conducted measurements show that the materials on the objects in the cabin have a higher reflectivity for IR light as compared to visible light. As a result, a higher received optical power is expected for an IR signal. Therefore, the study is focused on an optical cellular network with IR APs. Due to the complexity and specific nature of the problem, the MCRT simulation approach is employed. For a given transmitter model, the simulation outputs the irradiance distribution of the optical signal throughout the entire volume of the aircraft cabin for a specific receiver model with a variable photosensitive area. The path loss in particular volumes of the setup is then expressed as a function of distance for the given transmitter and receiver characteristics. The generally accepted exponential model for RF path loss is assumed and verified. This generic path loss model is employed in the estimation of the effective path loss in the aircraft cabin, including the specific properties of the transmitters and receivers, such as radiation/detection characteristics and FOV. The obtained path loss profiles allow for computation of SIR maps over cross-sections through the cabin. In this study, the SIR maps for the particular cellular division of the cabin are directly computed from the optical power distribution in the entire cabin volume.

Given a communication setup with highly reflective materials, it is expected to encounter LOS paths and NLOS paths with short-range single reflection or NLOS paths with long-range multiple reflections due to obstruction. These distinctive cases of the general behavior of the optical wireless signal propagation are the main objective of the MCRT irradiation simulation. They represent the majority of transmission scenarios

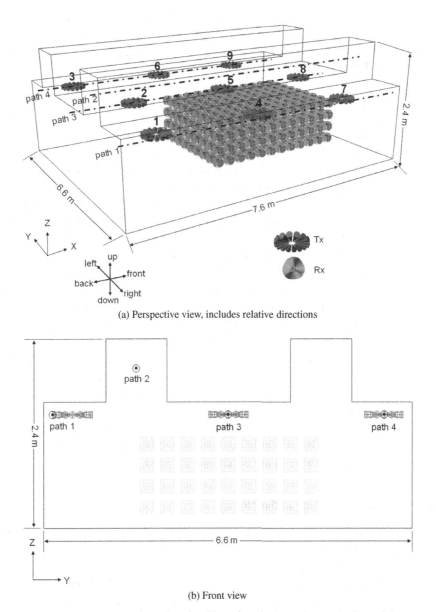

(a) Perspective view, includes relative directions

(b) Front view

Figure 2.8 Generalized model of the aircraft cabin [16]. It includes four paths for path loss estimation. NLOS path loss is estimated along paths 1 and 2. LOS path loss is estimated along paths 3 and 4. Nine transmitter positions are depicted, each transmitter consists of 16 LEDs. A simplified cuboid part of the receiver array is illustrated, each receiver consists of 6 PDs. Dimensions of the setup are included. The relative directions "front," "back," "up," "down," "left," and "right" are used for orientation in the cabin. ©2009 IEEE

in the volumes of the geometry for the chosen cellular design, and thereby allow for computation of SIR maps in similar environments and communication setups. The lines along which the path loss is estimated are shown in the generalized model of the aircraft cabin in Fig. 2.8. The LOS path loss is investigated along paths 3 and 4. A transmitter and the receivers on the specified paths have a direct LOS and FOV alignment. Path 3 represents an LOS condition enhanced by a single-reflection component coming from the ceiling. Path 4, in addition, collects the multiple reflections coming from the sidewall and the ceiling. The NLOS path loss is estimated along paths 1 and 2. Path 1 is located at the same height as transmitter 1, but it is closer to the sidewall. It represents an NLOS condition enhanced by a single-reflection multipath component coming from the sidewall. This particular scenario allows for examination of NLOS effects due to misalignment between transmitter and receiver in AZ, and short-range reflection. Path 2 is placed higher than transmitter 1 in the middle of the open volume in the neighborhood of the luggage compartments between the first and second row of transmitters. It is obstructed by a corner in the geometry, so that LOS is impossible. The optical power arrives at the receivers through multiple reflections. Hence, this scenario allows for the investigation of path loss due to transmitter and receiver misalignments in EL, long-range reflection, and obstruction.

Finally, SIR distribution maps in the aircraft cabin are calculated for wavelength reuse factors of 1, 2, and 3. The cell division of the setup and the wavelength assignment for the three reuse factors are presented in Fig. 2.9. The particular cellular structure is motivated by a potential application scenario inside the cabin. This is inter-system communication between the passenger service units (PSUs), which are located above the seats. If these units are connected wirelessly, a significant amount of cabling, and thus aircraft weight, can be saved. Moreover, a reduction in the cabling will enable a modular seat design, and seat layout reconfigurations are easier to accommodate. Therefore, in this study, it is assumed that the transceivers are integrated into the PSUs. In order to minimize the impact of the actual location of the optical AP on the variation of the SIR, an almost constant inter-site or inter-optical-access-point distance is maintained. In addition, the potential application scenario inside the aircraft cabin motivates the selection of the LOS and NLOS paths for path loss estimation.

2.6.3   Cabin geometry and materials

As a first step of the simulation modeling, the propagation environment of the optical wireless signal needs to be defined. The aircraft cabin constitutes a geometrically controlled environment, i.e. there are a finite number of the interior configurations for which the case of associated communication use can be defined, e.g. flight crew communication, flight entertainment, and aircraft maintenance on the ground. The geometric three-dimensional (3D) model of the aircraft cabin is constructed using the CAD tool Rhinoceros 3D [119]. The software provides a flexible and interactive modeling interface. A generalized model for the aircraft cabin is presented in Fig. 2.8. It serves as an example for the approximate dimensions and volumes in the considered aircraft section.

A major task is the definition of reflection coefficients of the materials on the different objects in the geometry. It is assumed that the materials primarily reflect in a diffuse fashion with a small specular portion. Since an IR emitter has a very narrow light spectrum, the focus of the simulation is on monochromatic light. Therefore, the attributes of the materials are defined through measured diffuse and specular reflection coefficients for light at 870 nm. There are a number of commercially available LEDs

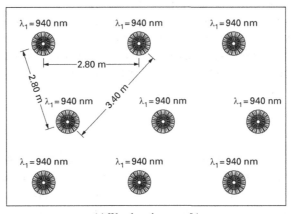

(a) Wavelength reuse of 1

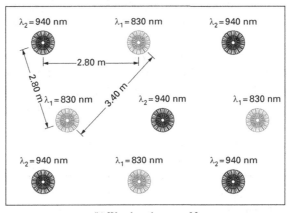

(b) Wavelength reuse of 2

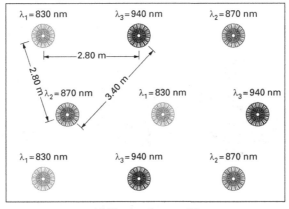

(c) Wavelength reuse of 3

Figure 2.9 Cell division and wavelength assignment for reuse factors of 1, 2, and 3 [17].
©2011 EurAAP

and PDs which have the peak of their spectral characteristics at this wavelength. The diffuse and specular reflection coefficients of the simulated materials are on average 0.85 and 0.1, respectively, and constitute a highly reflective environment. It is assumed that the materials with measured reflection characteristics cover most of the surfaces of the geometry, and the rest of the material types in the cabin do not differ significantly from the ones measured.

2.6.4   Access points

In order to simulate the light signal distribution in a cellular network, a transmitter model is defined. A single transmitter must cover the entire cell area, and it constitutes an AP. Furthermore, it needs to be as compact as possible. Hence, an omnidirectional radiation pattern is employed. Each transmitter is made out of 16 non-overlapping point light sources [102], shown in Fig. 2.10. The light sources are modeled according to the specifications of the Vishay IR emitter TSFF5210 [120], and form the desired close to omnidirectional radiation pattern. Another way to define the pattern is to employ fewer non-overlapping LEDs with broader radiation characteristic. However, the radiation intensity is not uniformly distributed over the entire FOV. It drops by half at the border of the specified FOV angle. Hence, a formation of the latter type does not support all directions with close to equal light intensity. An alternative approach is to consider overlapping radiation patterns, which can also yield an omnidirectional light distribution.

Input parameters of the LED are its radiation characteristic (goniogram), FOV of $\pm 10°$, and an optical power of 50 mW. As a result, the transmitter has a total optical power of 800 mW. Nine transmitters are positioned in the setup as shown in Fig. 2.8. The light signal power distribution over the entire setup volume is simulated distinctly for each transmitter. This enables the separation of source powers and thus allows for the determination of spatial SIR maps.

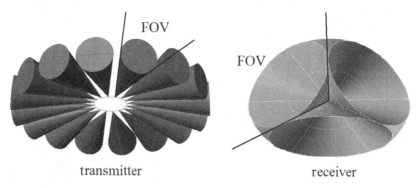

Figure 2.10 Omnidirectional radiation characteristic of a transmitter and a nearly isotropic detection pattern of a receiver [16]. The FOVs of an individual LED and a PD are illustrated. At the transmitter, 16 LEDs with FOV of $\pm 10°$ are placed around a circle with diameter of 2 cm. At the receiver, 6 PDs with FOV of $\pm 45°$ are arranged in a cubic formation to cover all 6 spatial directions. ©2009 IEEE

To estimate the path loss along predefined paths, the irradiance in the setup volume is sampled by means of a planar irradiance observer tool, which is a planar collection of receivers [102]. A receiver in the setup is modeled according to the specifications of the Vishay PD TESP5700 [121] with an FOV of $\pm45°$. For the calculation of a path loss example with the particular PD, the photosensitive area of 5.7 mm^2 is used. No collimator lens is utilized to concentrate the optical beam impinging on the photosensitive area of the flat-plate PD. Six PDs are placed together to form a single receiver as shown in Fig. 2.10. It covers all six spatial directions and allows for estimation of the path loss in a particular direction. With the help of equidistantly separated irradiance observer planes, a three-dimensional array of $200 \times 200 \times 100$ receivers is defined. Each observer plane is separated from its neighbor by approximately 3 cm in horizontal and vertical directions and has a resolution of approximately 3 cm $\times$ 4 cm per receiver area. Therefore, the maximum photosensitive area of a single PD is constrained up to 3 cm $\times$ 4 cm. A simplified part of the cuboid three-dimensional array of receivers is presented in Fig. 2.8.

Clearly, the selection of the LED and PD and the choice of transmitter and receiver models have an impact on the effective path loss in the setup. The photosensitive area of the PD scales the received optical intensity and constitutes a constant addend in the logarithmic domain. It is a variable in the proposed path loss model. The photosensitive area of the PD does not play a role in the calculation of the SIR maps, because when increasing the signal level with a larger photosensitive area, the interference level is proportionally enhanced as well. However, the FOV of the PD dictates which rays will be counted in the calculation of the irradiance distribution, depending on the angle of arrival. Thus, the chosen model of the receiver plays an important role in the path loss estimation methodology. The design makes it possible to separate the signal power coming from different directions and thus allows for distinction of the direct and multipath component arriving at the PD. The goniogram of the LED in combination with the reflection properties of the objects in the setup determines the power of the direct and multipath components arriving at each PD. Therefore, the radiation pattern of the transmitter has a major impact on the signal power distribution along a certain path. Since the simulation focuses not only on LOS, but also on NLOS path loss and SIR estimation, the omnidirectional radiation characteristic of the transmitter provides a high-power multipath signal component.

The limitations of the irradiation simulation are related to the stochastic accuracy and the abstraction level of the modeling approach. Of key relevance are the MCRT simulation stopping criterion of stochastic confidence and the definition of the simulation input parameters [122]. The geometric model of the aircraft cabin has a tessellation accuracy related to the resolution of the polygon meshes, which leads to small uncertainties in the MCRT calculation. The reflection characteristics of the materials in the aircraft cabin are defined through measured diffuse and specular reflection coefficients, which are subject to the tolerance of the measurement instruments and setup. Housing geometries of the LEDs and PDs are not considered. The LEDs are modeled as point sources with a radiation characteristic and FOV, whereas the PDs are modeled as point receivers with FOV.

2.6.5   Photobiological safety

The design of the transmitter is subject to eye safety regulations. According to the BS EN 62471:2008 standard for photobiological safety of lamps and lamp systems [44], non-coherent diffuse continuous-wave-modulated LEDs belong to the exempt group classification and pose no photobiological hazard for the human eye if the irradiance does not exceed 100 W/m^2 at a distance of 0.2 m from the optical source in the direction of maximal directivity within 1000 s. The setup for calculation of the human eye pupil irradiance under the requirements specified in the BS EN 62471:2008 standard [44] is illustrated in Fig. 2.11. It shows an LED, placed in the XY-plane and at the origin of a right-hand spherical coordinate system with radius r, zenith angle θ, and azimuth angle ϕ. The LED is directed orthogonally to the cross-section of a human eye pupil. The axis of symmetry for both the LED and the eye pupil is in the Z direction. The FOV semi-angle of the LED is given by $\theta_{FOV,Tx}$, and the angle subtended by the eye pupil, and the origin is denoted by θ_{eye}. The distance between the LED and the eye is denoted by d, r_{eye} is the radius of the eye pupil, and r_{FOV} is the distance made by the angle $\theta_{FOV,Tx}$ when crossing the plane through the eye pupil and the Z-axis.

The distance r_{FOV} is expressed as follows:

$$r_{FOV} = d \tan(\theta_{FOV,Tx}) . \tag{2.24}$$

Applying the values for the specified distance, d, of 200 mm and the FOV semi-angle of the LED, $\theta_{FOV,Tx}$, of 10° in (2.24), a value for r_{FOV} of 35.3 mm is obtained, which is larger than the radius of the average eye pupil, r_{eye}, of 2 mm in day light [123]. Therefore, the eye pupil subtends an angle which is much smaller than the FOV of a single LED. Since, in addition, the design of the radiation pattern of the transmitter does not involve the overlapping of the radiation patterns of the individual LEDs, it is safe to assume that only one LED is responsible for the radiation in the direction of maximal directivity, and thus irradiating the eye.

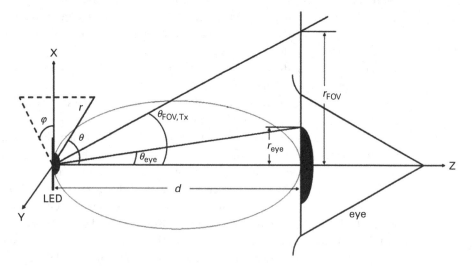

Figure 2.11 Setup for calculation of the irradiance on the human eye from the IR optical source [16]. ©2009 IEEE

As a next step, the portion of the optical power emitted by the LED that falls onto the eye pupil is calculated. For this purpose, the mathematical description for the radiation pattern of a generalized Lambertian source [3] is used as follows:

$$R_{LED}(\theta, n_{Tx}) = P_T \frac{n_{Tx} + 1}{2\pi} \cos^{n_{Tx}}(\theta) . \tag{2.25}$$

Here, the mode number of the radiation lobe, n_{Tx}, can be expressed as a function of the FOV semi-angle, $\theta_{FOV,Tx}$, as follows:

$$n_{Tx} = \frac{\log(1/2)}{\log\left(\cos(\theta_{FOV,Tx})\right)}. \tag{2.26}$$

The amount of the emitted power irradiating the eye pupil, P_{eye}, can be expressed as:

$$P_{eye} = \int_0^{2\pi} \int_0^{\theta_{eye}} R_{LED}(\theta, n_{Tx}) \sin(\theta) \, d\theta \, d\phi$$

$$= P_T \left(1 - \cos^{n_{Tx}+1}(\theta_{eye})\right) , \tag{2.27}$$

where

$$\theta_{eye} = \arctan\left(\frac{r_{eye}}{d}\right) . \tag{2.28}$$

The irradiance of the human eye, E_{eye}, is then calculated as follows:

$$E_{eye} = \frac{P_{eye}}{\pi r_{eye}^2} . \tag{2.29}$$

Substituting (2.26) and (2.28) into (2.27), then (2.27) into (2.29), and applying the total optical power of the LED, P_T, of 50 mW, the FOV semi-angle of the LED, $\theta_{FOV,Tx}$, of 10°, the radius of the eye pupil, r_{eye}, of 2 mm, and the distance, d, of 200 mm yield an irradiance value of 9.2 W/m². This value is much less than the specified 100 W/m² in the BS EN 62471:2008 standard [44] and, therefore, the eye safety regulations are fulfilled for the designed transmitter.

2.6.6 Estimation of line-of-sight path loss and shadowing

The irradiance distribution in the setup with the previously described optical communication scenario is simulated up to an MCRT simulation stopping criterion of 3% stochastic confidence. In order to evaluate the path loss along the four selected paths in the aircraft cabin, the path loss exponent, ζ, and the standard deviation of the shadowing component, σ_{shad}, have to be estimated. Since the term with the photosensitive area of the PD is only an addend in the equation, it only shifts the path loss curve vertically, without changing its slope or the standard deviation of the scattering due to shadowing effects. Specifically, the photosensitive area of the Vishay PD is used in order to estimate the values of the constants in the equation for the four paths. Accordingly, the estimated values can be applied to the more general path loss model in (2.15). The results for the four paths are presented in Table 2.1.

The LOS path loss is estimated along paths 3 and 4. For reasons of consistency, it is important to refer to the relative directions shown in Fig. 2.8(a). Path 3 represents a

Table 2.1 Estimated parameters for the path loss model (2.15) on the four paths in the aircraft cabin

Paths	d_{ref}[m]	P_T[mW]	$E(d_{ref})$[mW/m²]	ζ	σ_{shad}[dB]
Path 1 – NLOS enhanced by a single reflection	0.42	800	611.24	2.26	1.27
Path 2 – NLOS enhanced by multiple reflections	1.45	800	53.11	1.28	0.7
Path 3 – LOS enhanced by a single reflection	0.04	800	280036.82	1.94	0.57
Path 4 – LOS enhanced by multiple reflections	0.04	800	280682.37	1.92	0.81

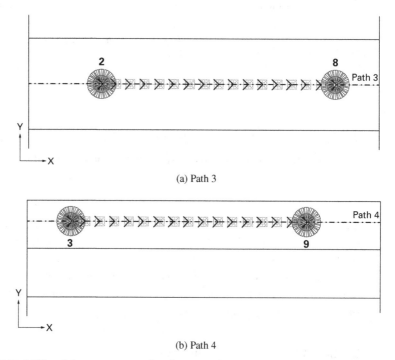

(a) Path 3

(b) Path 4

Figure 2.12 LOS path loss exponent estimation scenarios along path 3 and path 4 [16]. ©2009 IEEE

direct LOS communication scenario with FOV alignment between the single operational transmitter 2 and the receivers along the path, which is enhanced by the single-reflection component coming from the baggage compartment. Therefore, the path loss exponent along the path is estimated when only the signal arriving at the receivers in the "back" receive direction is counted. The procedure is consistently repeated along path 4 and transmitter 3. An illustration is provided in Fig. 2.12. The same receive direction is considered for LOS shadowing component estimation. Even though no dynamic variation of the environment is simulated and the path loss results are obtained through computation on a static setup, it is possible to approximate the standard deviation of the LOS shadowing component on paths 3 and 4. Thus, receivers within the FOV of the

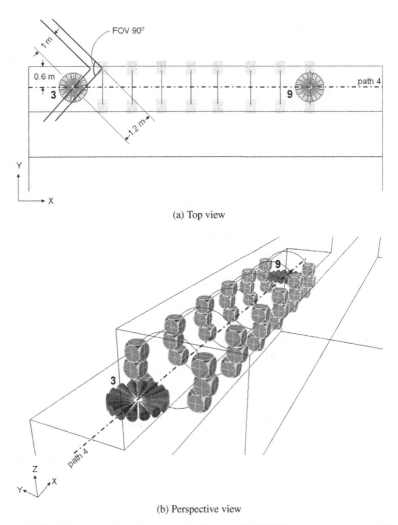

(a) Top view

(b) Perspective view

Figure 2.13 LOS shadowing estimation scenario along path 4 [16]. Receivers are selected from circles around the path. They fall within the FOV of transmitter 3 and are located at distance from 1.2 m to 5.6 m away from it. Some of the circles are illustrated. ©2009 IEEE

respective transmitter are selected from circles which are orthogonal to and centered on the respective path. A simplified illustration is shown in Fig. 2.13. The selected receivers on each circle are equidistant to the transmitter and have the same FOV orientation towards it. The radius of the circles on path 4 is chosen to be 0.6 m in order to include the receivers which are closest to the sidewall and experience the highest shadowing. For consistency, the same radius is chosen for the circles around path 3. In this configuration, an LOS signal component is received at distances from the transmitter greater than 1 m. To minimize the effect of the FOV on the LOS path loss, receivers located at distances greater than 1.2 m are considered in the estimation of the LOS shadowing, as given in Fig. 2.13(a).

Since transmitters 2 and 3 are positioned on paths 3 and 4, respectively, the LOS path loss is estimated for distances from 0.04 m to 5.6 m away from the respective operational transmitter. The two resulting path loss graphs are presented in Figs. 2.14(a) and 2.14(b). The path loss exponent, ζ, is the slope of the path loss graph, obtained after linear regression. Also, the irradiance distribution in a vertical cross-section of the setup along the two paths is visualized. It is evident that the direct LOS path loss graph exhibits linear behavior in the log domain. Therefore, the initial assumption that the general RF path loss equation can be employed for the description of the optical wireless path loss is verified. The minor difference between the path loss exponents on path 3 and path 4 in Table 2.1 can be attributed to the stochastic error due to the MCRT simulation stopping criterion, rather than the different signal components arriving at the receivers along the paths. A notable observation is that the single-reflection multipath component coming from the wall along path 4 does not have much influence on the path loss exponent in comparison to path 3 because of the strong direct LOS component.

According to the described LOS shadowing estimation methodology, the two resulting path loss scattering plots along paths 3 and 4, and the corresponding probability density functions are presented in Figs. 2.15(a) and 2.15(b), respectively. The probability density function is calculated from the variation of the scattered path loss values around a linear regression mean curve. The results suggest that shadowing can be modeled as a log-normally distributed random variable in the linear domain, which corresponds to a zero-mean Gaussian distribution in the logarithmic domain. Hence, the standard deviation of the shadowing component, σ_{shad}, around paths 3 and 4 is evaluated and presented in Table 2.1. Because of the objects in proximity, e.g. the curved shapes of the sidewall panel and the light panel, path 4 exhibits a higher shadowing standard deviation and more outliers in the tails of the probability density function, as compared to path 3. In addition, it is shown in Fig. 2.15(a) and Fig. 2.15(b) that at shorter distances the path loss points are more scattered around the linear regression mean curve. This effect can be attributed to the alignment between the FOVs of the transmitter and the respective receivers along the path. With the increase of distance, a better alignment is achieved, the single-reflection multipath component becomes much less intensive than the LOS component, and as a consequence the scattered points deviate less from the linear regression mean curve.

2.6.7   Estimation of non-line-of-sight path loss and shadowing

The NLOS path loss is estimated along paths 1 and 2. Path 1 represents an NLOS condition due to misalignment of the operational transmitter 1 and the receivers along the path in AZ, which is enhanced by the short-range single-reflection component coming from the sidewall. The distance between transmitter 1 and path 2 is 0.42 m, whereas the path is located 0.18 m away from the sidewall. Consequently, to estimate the path loss exponent only, the signal arriving at the receivers along the path in the "right" receive direction is considered as shown in Fig. 2.16(a). This receive direction is chosen when estimating the NLOS shadowing along the path. Thus, receivers on concentric circles

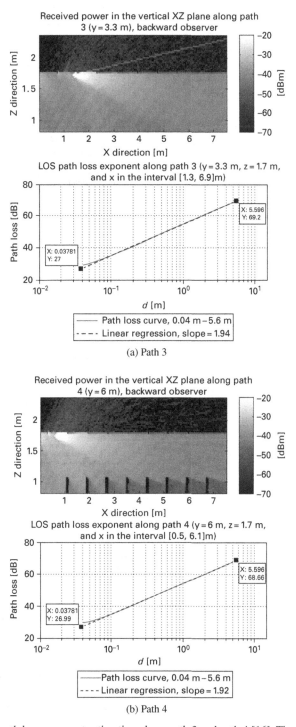

Figure 2.14 LOS path loss exponent estimation along path 3 and path 4 [16]. The distance from the transmitter is denoted by d. ©2009 IEEE

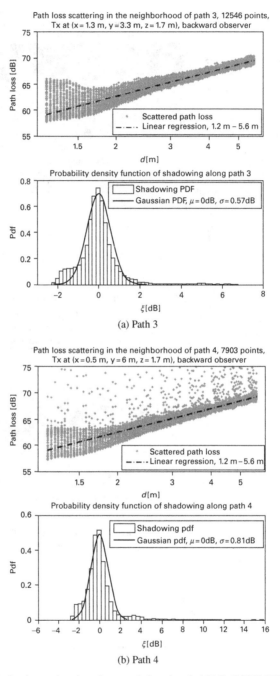

Figure 2.15 LOS shadowing estimation along path 3 and path 4 [16]. ©2009 IEEE

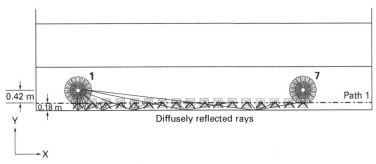

(a) Path 1. Only the signal reflected by the sidewall is counted. The distances between transmitter 1, path 1, and the sidewall are included

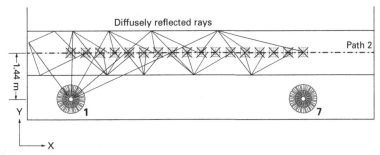

(b) Path 2 is above the locations of transmitters 1 and 7. The receivers along the path receive the multiply reflected multipath components of the signal in the "front," "back," "left," and "right" receive directions. The distance between transmitter 1 and path 2 is included

Figure 2.16 NLOS path loss exponent estimation scenarios along path 1 and path 2 [16]. ©2009 IEEE

around the projected position of the transmitter onto the vertical plane along the path are selected, as shown in Fig. 2.17.

The NLOS path loss scenario along path 2 studies the impact of obstruction, as well as transmitter and receiver misalignments in EL as opposed to AZ (as in all previous cases). The scenario is illustrated in Fig. 2.16(b). The NLOS conditions are induced by the corner in the geometry between the path and the operational transmitter 1. The direct (shortest) distance between transmitter 1 and path 2 amounts to 1.45 m. The receivers along the path are irradiated by the multiple reflections multipath signal component. Thus, in order to estimate the path loss exponent, the receivers along the path collect the signal that impinges on the "back," "front," "left," and "right" receive directions. To estimate the shadowing component, receivers in the specified directions are selected from concentric spheres around the position of transmitter 1. Furthermore, only those receivers on the spheres are counted, which are located in the small cuboid volume in the neighborhood of path 2.

Because of the horizontal displacement between transmitter 1 and path 1, the NLOS path loss is estimated for distances from 0.42 m to 5.61 m away from the transmitter. Similarly, due to horizontal and vertical displacement between transmitter 1 and path 2,

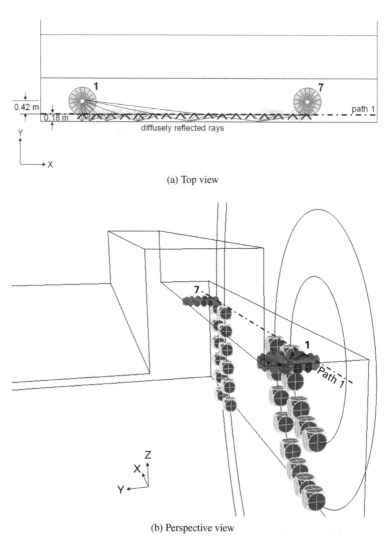

(a) Top view

(b) Perspective view

Figure 2.17 NLOS shadowing estimation scenario along path 1 [16]. Receivers are selected from concentric circles in the vertical plane along path 1 around the orthogonally projected position of transmitter 1 onto this vertical plane. The receivers are located at distances from 0.4 m to 5.6 m away from the transmitter. Some of the circles are illustrated. ©2009 IEEE

the NLOS path loss along path 2 is estimated for distances from 1.45 m to 5.78 m. The resulting path loss graphs for the two NLOS scenarios are presented in Figs. 2.18(a) and 2.18(b), respectively. Accordingly, the path loss scattering plots and the probability density function of the shadowing components along path 1 and path 2 are displayed in Figs. 2.19(a) and 2.19(b). The resulting values of the NLOS path loss exponents and the standard deviation of the shadowing components along the two paths are given in Table 2.1. An interesting observation from the two presented NLOS scenarios is that the path loss graphs in Figs. 2.18(a) and 2.18(b) intersect at a distance between 4 m and 5 m. This means that the directed NLOS scenario along path 1 benefits from the strong

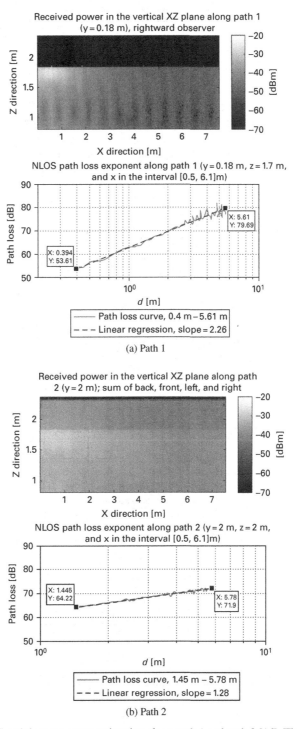

(a) Path 1

(b) Path 2

Figure 2.18 NLOS path loss exponent estimation along path 1 and path 2 [16]. The distance from the transmitter is denoted by d. ©2009 IEEE

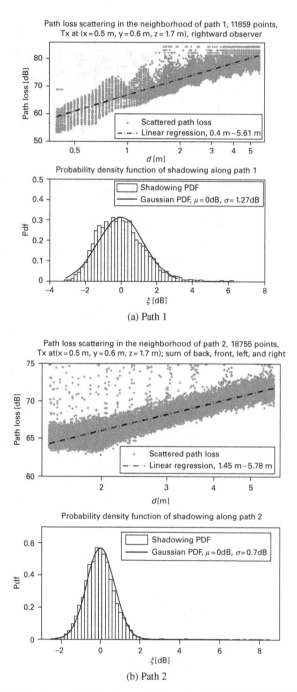

(a) Path 1

(b) Path 2

Figure 2.19 NLOS shadowing estimation along path 1 and path 2 [16]. ©2009 IEEE

single reflection at short to medium distances, whereas at larger distances the power of the single strong reflection at the receiver decreases more rapidly than the power of the sum of the multiple reflections in the full diffuse NLOS case along path 2. This is because the single-reflection propagation is limited at larger distances by the considered FOV misalignment between transmitter and receiver. Similar to the LOS scenarios, the objects very close to path 1 increase the standard deviation of the shadowing component. On the contrary, path 2 is more distant from the neighboring objects, e.g. the two luggage compartment rows, and the standard deviation of the log-normal shadowing is therefore less pronounced. In the case of path 1, the sidewall panel topology casts shadows onto the receivers selected from the vertical plane in immediate proximity, and this results in more outliers in the tails of the corresponding probability density function.

The path loss estimation scenarios demonstrate that NLOS in the aircraft cabin degrades the effective path loss by approximately 10 dB as compared to the LOS scenarios. However, using the path loss parameters determined in this study, respective link budget margins could be considered in the optical wireless system design.

2.6.8   Signal-to-interference ratio maps

The previously described division of the aircraft cabin into nine cells facilitates the estimation of the SIR distribution for different wavelength reuse factors. In the simulation, it is assumed that the interference is merely caused by co-channel signals, and the adjacent channel interference (ACI) can be controlled by proper selection of LED-PD pairs. An example can be given for two LED-PD pairs, offered by Vishay Semiconductors. The LED TSFF5210 and the PD TESP5700 have peak spectral emission and responsivity at 870 nm. The LED VSMB2020X01 [124] and the PD BPW41N [125] have peak spectral emission/sensitivity at 940 nm. Their 3-dB bandwidths around the peak wavelengths do not overlap, and therefore ACI can be assumed to be sufficiently low. Since the light distribution for each of the nine transmitters is simulated separately, it is convenient to identify the interfering cells for every intended cell, according to the wavelength division and assignment in Fig. 2.9.

The positioning of the transmitters is suitable for receiver locations in the neighborhood of the horizontal plane in which the transmitters are placed. However, given the mobility of users and the different cases of use, it is difficult for a practical transmitter to radiate the same power in all possible elevation directions. Therefore, the SIR distributions at different horizontal levels through the cabin provide an important insight into the achievable SIR when the transmitters do not point in the direction of the actual location of a user, i.e. the SIR when the user is located at the boundary or even outside the FOV of a transmitter. SIR distribution levels at 0.5 m and 1 m below the transmitters' plane are investigated. The resulting maps are presented in Figs. 2.20, 2.21, and 2.22 for wavelength reuse factors of 1, 2, and 3, respectively. The few cross-formed areas with higher intensity noticeable on the SIR maps represent artifacts due to the receiver model in the simulation and should be ignored.

The SIR values at the extreme positions in the optical wireless cellular network, i.e. in the center of the cell and at the border of the cell, are summarized in Table 2.2 for

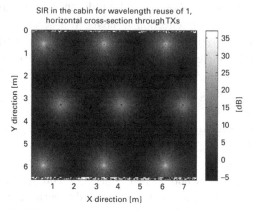

(a) SIR in the horizontal plane passing through the transmitters

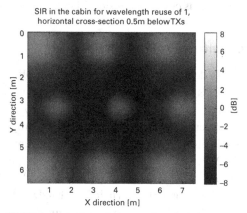

(b) SIR in the horizontal plane at distance 0.5 m below the transmitters

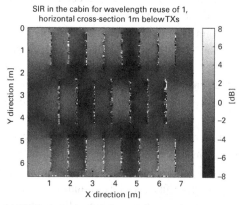

(c) SIR in the horizontal plane at distance 1 m below the transmitters

Figure 2.20 SIR distribution for wavelength reuse of 1 [16]. ©2009 IEEE

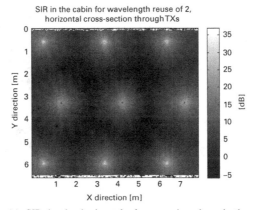

(a) SIR in the horizontal plane passing through the transmitters

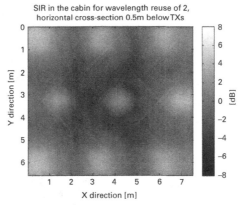

(b) SIR in the horizontal plane at distance 0.5 m below the transmitters

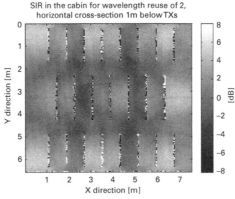

(c) SIR in the horizontal plane at distance 1 m below the transmitters

Figure 2.21 SIR distribution for wavelength reuse of 2 [16]. ©2009 IEEE

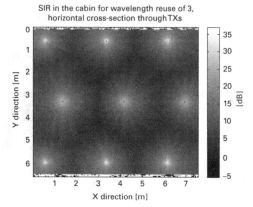

(a) SIR in the horizontal plane passing through the transmitters

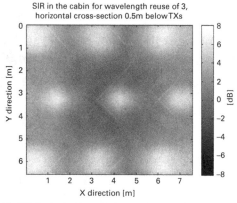

(b) SIR in the horizontal plane at distance 0.5 m below the transmitters

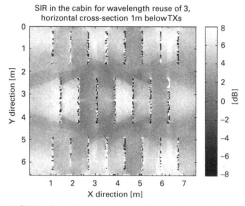

(c) SIR in the horizontal plane at distance 1 m below the transmitters

Figure 2.22 SIR distribution for wavelength reuse of 3 [16]. ©2009 IEEE

Table 2.2 SIR distribution in the three cross-sections through the aircraft cabin

Wavelength reuse		Reuse of 1		Reuse of 2		Reuse of 3	
Number of interfering cells		8		3 or 4		2	
SIR plane	SIR position	simul.	calc.	simul.	calc.	simul.	calc.
Cross-section in the Tx plane	SIR in the cell center [dB]	30	28.97	35	34.41	37	37.45
	SIR at the cell border [dB]	−5.5	−5.78	−2	−2.07	3	3.51
Cross-section 0.5 m below the Tx plane	SIR in the cell center [dB]	1		5		7.2	
	SIR at the cell border [dB]	−8		−4.3		−1.5	
Cross-section 1 m below the Tx plane	SIR in the cell border [dB]	2		5.9		8	
	SIR at the cell center [dB]	−7		−3.4		0.4	

all the specified topologies. The SIR in the center of the cell is defined as the SIR at the PD which is closest to the respective transmitter. The shortest distance, in this case, is 0.04 m due to the dimensions of the receiver array and the relative position of the transmitter. Accordingly, the border of the cell is defined as the PD in the corner between three cells.

A study conducted in [63] suggests that an optical wireless network with a hexagonal cell configuration and a channel reuse of 3 is CCI-limited, i.e. the CCI signal is greater or comparable to the desired signal at the border of the cell, for cell radii less than 2 m. Since reuse factors of 1 and 2 provide a higher number of interfering cells in a hexagonal cell configuration, the CCI limitation is even more pronounced. The results obtained for the designed cellular network in the aircraft cabin demonstrate that this statement is valid for the simulated environment. The SIR at the extreme positions in the network is evaluated in order to assess the actual quality of the link in this environment. Furthermore, the SIR in the chosen deployment scenario, the PSU communication, is verified by a calculation with the single-reflection LOS path loss estimation parameters in Table 2.1 and (2.15). For this purpose, one of the nine cells is selected and the SIR values at its center and border are computed. The transmitter in this cell is located at 3.35 m in the X direction and at 0.59 m in the Y direction. The closest PD, i.e. the center of the irradiance distribution in the cell, is placed at 3.38 m in the X direction and at 0.57 m in the Y direction. The PD in the corner between the intended cell and its two neighboring cells defines the border of the intended cell at 2.74 m in the X direction and at 2.32 m in the Y direction. According to the wavelength assignment in Fig. 2.9, 8 interfering transmitters are identified for wavelength reuse of 1, 3 interfering transmitters for wavelength reuse of 2, and only 2 interfering transmitters for wavelength reuse of 3. The distances from the interfering transmitters to the center and border points of the intended cell are presented in Table 2.3. Applying these distances to the path loss

Table 2.3 Calculation of the SIR in the cross-section through the transmitters' plane for the cell at (3.35 m, 0.59 m)

Wavelength reuse	Reuse of 1		Reuse of 2		Reuse of 3	
SIR position	center	border	center	border	center	border
Distances to	2.77	1.69	3.38	1.69	4.48	3.69
interfering	2.81	1.7	4.48	3.69	5.39	4.32
transmitters [m]	2.83	2.79	5.39	4.32		
	3.38	3.69				
	4.48	3.82				
	5.39	4.25				
	6.06	4.32				
	6.09	4.99				
Distance to intended transmitter [m]	0.04	1.83	0.04	1.83	0.04	1.83
Interference signal [dBm]	−26.94	−24.4	−32.38	−28.11	−35.42	−33.69
Intended signal [dBm]	2.03	−30.18	2.03	−30.18	2.03	−30.18
SIR [dB]	28.97	− 5.78	34.41	− 2.07	37.45	3.51

model (2.15) and selecting the single-reflection LOS path loss estimation parameters in Table 2.1, the respective path losses from the interference cells to the center and points of the intended cell are computed for the three wavelength reuse cases. As a result, the respective total interference signals at the specified points are obtained. Having the distance from the intended transmitter to the center and border of its cell, the intended signal values are used for the calculation of the SIR in Table 2.3. The calculated mean values in the center and at the border of the cell are also given next to the ones extracted from the SIR distribution in Table 2.2 for the cross-section through the transmitters. The values obtained from the two approaches closely match. Therefore, the single-reflection LOS path loss model is a suitable representative of the SIR distribution through the transmitters' plane. It can be used for SIR calculation in similar setups.

In general, the results demonstrate that lower wavelength reuse factors, e.g. reuse of 1 and 2, cannot provide sufficient SIR at the border of the cell in the described network. In such cases, communication cannot be supported at the cell boundaries without some form of intelligent resource allocation such as dynamic channel allocation, fractional channel reuse, or soft reuse. When assuming static resource allocation techniques, higher reuse factors are needed to enhance the SIR distribution in the aircraft cabin. An SIR which exceeds 3 dB at the border of a cell is achieved for wavelength reuse of 3 or greater. This guarantees a minimal level of data throughput for audio and video broadcast at a practical BER of 10^{-7}, using, for example, non-recursive convolutional codes [126], turbo codes, or low density parity check (LDPC) codes. Nonetheless, higher wavelength reuse factors can be considered if the favorable optical spectrum, over which the materials in the environment exhibit admissible reflection, is wide enough to accommodate all optical bands.

Comparing Figs. 2.20, 2.21, and 2.22 a non-intuitive observation can be made: the SIR in the cross-section at distance of 1 m is higher than the SIR at half the distance at

0.5 m from the transmitters' plane. This behavior can be attributed to the fact that the seats in the aircraft cabin have a shadowing effect on the interfering signal. This means that shadowing effects could be exploited constructively for static or quasi-static transmission scenarios. In a different scenario, when the main path in the highly reflective setup is temporarily blocked by a moving passenger, a secondary path reflected by a nearby object in the cabin should be able to provide sufficient energy for data detection. Alternatively, the user has to be served by a neighboring optical AP. Again, this would require dynamic channel assignment algorithms in such a network.

2.7 Summary

In this chapter, the fundamentals of an OWC setup have been discussed, including the front-end design, the geometry of the wireless link, and the resulting optical wireless channel. The typical building blocks of the transmitter and the receiver have been presented. In addition, the LOS and NLOS communication scenarios have been introduced which can be combined to enhance the irradiance at the receiver and the coverage in given indoor IR and VLC setups. A deterministic and a statistical model for the path loss of the optical wireless channel have been presented, including a summary of the mathematical details of the common ray-tracing techniques, such as the deterministic algorithm and MCRT.

A case study has been performed to determine the path loss of a cellular OWC system inside an aircraft cabin through a comprehensive MCRT irradiation simulation. Measured reflection coefficients of the materials in the cabin have been used for simulation modeling. This approach allowed for the characterization of propagation paths, without having to resort to expensive and time-consuming measurements. Moreover, spatial SIR maps of a particular deployment scenario have been calculated for different wavelength reuse factors. The OWC network has proven to be interference-limited, and the SIR at the border of the cells has shown to be maximized by higher wavelength reuse factors.

Furthermore, realistic LOS and NLOS channel models have been modeled, including path loss and log-normal shadowing. Specifically, the path loss exponent and the standard deviation of shadowing have been determined for particular LOS and NLOS scenarios. It has been demonstrated that even in the presence of a multipath component, the LOS optical path loss was dominated by the direct component, and it varied between 27 dB and 69 dB with a path loss exponent of 1.92 at distances from 0.04 m to 5.6 m. The considered NLOS scenarios investigated the single short-range reflection due to misalignment of transmitter and receiver in AZ and multiple long-range reflections due to obstruction. It has been shown that the former case experienced a larger path loss exponent of 2.26, resulting in path losses between 54 dB and 80 dB at distances from 0.4 m to 5.61 m. The latter scenario suffered a signal power attenuation with a lower path loss exponent of 1.28 and path losses between 64 dB and 72 dB at distances from 1.45 m to 5.78 m.

The results demonstrated that NLOS in an aircraft cabin degraded the performance by approximately 10 dB as compared to LOS, while the two became comparable at larger distances. Nonetheless, the level of degradation would still enable NLOS communication links also at shorter distances if the respective link budget margins could be accommodated. Also, the path loss fluctuations due to shadowing were not significant and robust communication links could be expected. The presented SIR maps demonstrated that in the studied deployment scenario an SIR of about 3 dB was obtained at the edge of the cells for wavelength reuse of 3. This would guarantee a minimal level of data throughput at a practical BER. Consequently, for the deployment of full wavelength reuse systems, interference avoidance or interference mitigation techniques such as interference-aware resource allocation and scheduling were required [127–130].

The obtained path loss parameters, in combination with the SIR maps, could be used to calculate a realistic link budget, i.e. to determine the number of LEDs in a transmitter and PDs in a receiver, to design the cellular system, and to develop handover algorithms. As a result, this simulation platform provided the means for optical wireless system design, test of performance, and optimization.

3 Front-end non-linearity

3.1 Introduction

Because of the p-n junction barrier and the saturation effect of the light emitting diode (LED), the transmitter front-end has a limited linear dynamic range. Single-carrier pulse modulated signals such as multi-level pulse position modulation (M-PPM) and multi-level pulse amplitude modulation (M-PAM) have a probability density function (PDF) of their intensity levels with finite support of values. As a result, these signals can be conditioned within the limited linear dynamic range and transmitted with negligible non-linear distortion. Multi-carrier signals based on optical orthogonal frequency division multiplexing (O-OFDM), such as direct-current-biased O-OFDM (DCO-OFDM) and asymmetrically clipped O-OFDM (ACO-OFDM) with multi-level quadrature amplitude modulation (M-QAM), follow Gaussian and half-Gaussian distributions of the intensity levels for a large number of subcarriers. As a consequence, such signals have a high peak-to-average power ratio (PAPR), and they are transferred by the optical front-end in a non-linear fashion. This results in an increased bit-error rate (BER) or an increased electrical signal-to-noise ratio (SNR) requirement due to non-linear signal distortion. In order to alleviate this issue, signal pre-distortion with the inverse of the non-linear transfer function can be employed. The linear dynamic range is maximized by pre-distortion, and a linear signal transfer is obtainable between levels of minimum and maximum radiated optical power. As a result, the single-carrier pulse modulation signals are transferred without non-linear distortion and with increased electrical power efficiency. The non-linear distortion of the high PAPR O-OFDM signals is reduced to double-sided signal clipping [131, 132], and these signals also benefit from increased electrical power efficiency.

In this chapter, a generalized piecewise polynomial model is presented for the non-linear distortion function of the transmitter front-end [133]. It is a flexible and accurate representation of transmitter non-linearity. Furthermore, it enables signal pre-distortion with the inverse of the non-linear transfer function for the objective of maximizing the linear dynamic range of the optical front-end. The non-linear distortion of an information-carrying subcarrier in O-OFDM is modeled as a real-valued attenuation factor and additive non-linear noise component with a zero-mean complex-valued Gaussian distribution. The non-linear distortion parameters, such as the attenuation factor and the variance of the non-linear noise, are derived in closed form for the generalized non-linear transfer function and the double-sided signal clipping after pre-distortion [134].

3.2   Generalized non-linear transfer function

A light emitting diode (LED) is biased by a constant current source which supports
the entire range of forward voltages across the LED. The bias current is added to the
data-carrying current, yielding the forward current through the LED, I_f. The non-linear
transfer characteristic of the LED can be described by the non-linear relation between
the forward current, I_f, and the forward voltage, V_f, which can be translated into a non-
linear relation between the dissipated electrical power, P_{elec}, and the radiated optical
power, P_{opt}, as illustrated in Fig. 3.1. In this study, non-linearity is generalized as a
relation between the input current, $I_{in} = x$, linearly proportional to the square root
of the dissipated electrical power, P_{elec}, and the output current, $I_{out} = F(x)$, linearly
proportional to the radiated optical power, P_{opt}. Therefore, the large-PAPR current
signal, x, is subjected to a non-linear distortion function, $F(x)$, as it is passed through
the front-end block. The current signal, x, is directly related to the information-carrying
symbols, s, of the different modulation formats. For example, in M-PPM $x = s$, while
in M-PAM, M-QAM DCO-OFDM, and M-QAM ACO-OFDM $x = s + \beta_{DC}$, where β_{DC}
is the DC bias. Together with the signal variance, σ^2, β_{DC} is a control parameter that
can be used to optimally condition the symbols within the positive dynamic range, in
order to minimize the non-linear symbol distortion. The formulation of the optimization
problem is presented in Chapter 5. For the purpose of generality, the non-linear transfer
function of the transmitter front-end, $F(\cdot)$, is normalized, and the normalized non-linear
transfer function, $\Psi(\cdot)$, is defined as follows: $\Psi(s) = F(x) - F(\beta_{DC})$. The following

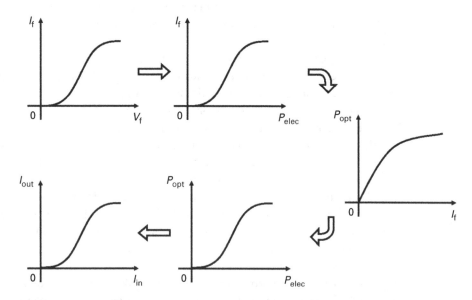

Figure 3.1 Typical non-linear transfer characteristic of an LED [133]. It includes the relations
between the forward current, I_f, the forward voltage, V_f, the radiated optical power, P_{opt}, and the
dissipated electrical power, P_{elec}. The model is generalized as a non-linear relation between the
input and output information-carrying currents, I_{in} and I_{out}. ©2013 IEEE

piecewise polynomial model for $\Psi(\cdot)$ is used as an accurate, flexible, and generalized representation of the normalized non-linear transfer function [133]:

$$\Psi(s) = \begin{cases} \sigma\lambda_1, & \text{if } s \leq \sigma\lambda_1, \\ \psi_1(s), & \text{if } \sigma\lambda_1 < s \leq \sigma\lambda_2, \\ \vdots & \\ \psi_{J-1}(s), & \text{if } \sigma\lambda_{J-1} < s \leq \sigma\lambda_J, \\ \sigma\lambda_J, & \text{if } s > \sigma\lambda_J. \end{cases} \tag{3.1}$$

Here, $\psi_j(s), j = 1, \ldots, J - 1$, are polynomial functions of the non-negative integer order n_j, and σ is the signal standard deviation. This is a flexible yet general way to represent any non-linear function. In addition, it facilitates the pre-distortion of the transmitted signal in order to linearize the dynamic range of the transmitter and to reduce the non-linear distortion to double-sided signal clipping. Here, $\lambda_{\hat{j}}, \hat{j} = 1, \ldots, J$, are real-valued normalized clipping levels that denote the end points of the polynomial functions, $\psi_j(s)$. They are relative to the standard form of the signal distribution. For example, in O-OFDM the clipping levels are relative to a standard normal distribution with zero-mean and unity variance. In DCO-OFDM, the levels can be positive as well as negative, while in ACO-OFDM the levels are strictly non-negative because of the non-negative half-Gaussian signal distribution.

3.3   Pre-distortion

Conventionally, the dynamic range of the transmitter to be modulated by the information-carrying signal is chosen as the portion of the transfer characteristic which can be considered as nearly linear. However, through signal pre-distortion with the inverse of the non-linear transfer function, the linear dynamic range of the transmitter can be maximized. As a result, the electrical alternating current (AC) power of the signal can be maximized at the transmitter in order to increase the electrical SNR at the receiver. This is illustrated in Fig. 3.2. A pre-distortion of the $V_f - I_f$ transfer characteristic of the optical transmitter has been considered in [135]. In order to generalize the non-linearity model, the pre-distortion of the non-linear $I_{in} - I_{out}$ transfer characteristic, i.e. $F(x)$, which is a superposition of the non-linear $V_f - I_f$ and $I_f - P_{opt}$ transfer characteristics, is considered in this study.

The piecewise polynomial function, $\Psi(s)$, can be applied to model any non-linear transfer function, $F(x)$, as follows: $\Psi(s) = F(x) - F(\beta_{DC})$. For a general non-linear sigmoid function, $F(x) = \Xi(x)$, non-linear signal distortion can be mitigated by pre-distortion of the signal with the inverse function $\Xi^{-1}(x)$. Two examples for the transfer function of an LED are presented in Fig. 3.3, one for a non-linear sigmoid function, $F(x) = \Xi(x)$, and one for a linearized function, $F(x) = \Phi(x)$. Due to the potential barrier in the p-n junction and the saturation effect of the LED, a linear transfer of the pre-distorted signal is obtainable only between points of positive minimum and maximum

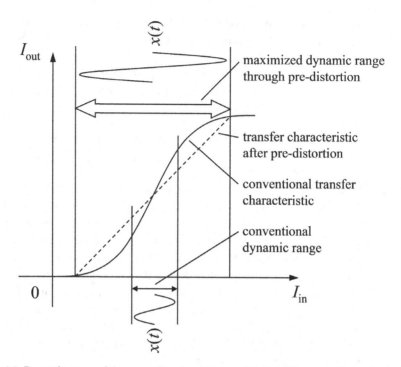

Figure 3.2 Dynamic range of the transmitter for signal modulation. The nearly linear dynamic range of a conventional non-linear transfer characteristic is illustrated, as well as the maximized linear dynamic range after signal pre-distortion

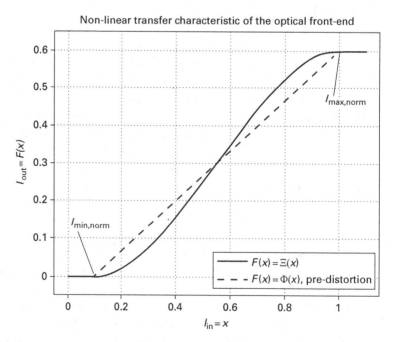

Figure 3.3 Non-linear transfer characteristic of the considered optical front-end [133]. The generalized piecewise polynomial function is denoted by $\Xi(x)$. The linearized characteristic after pre-distortion is denoted by $\Phi(x)$. ©2013 IEEE

input current, I_{min} and I_{max}, resulting in a limited dynamic range of transmitter front-end. Using pre-distortion, a linear relation is established between the radiated optical power and the information-carrying input current over the limited dynamic range. Without loss of generality, the input current is normalized to I_{max}, and the radiated optical power is normalized to P_{max}. As a result, the following normalized quantities are assumed for the boundaries of the dynamic range of the front-end in terms of normalized minimum and maximum optical power and current: $P_{min,norm} = I_{min,norm} = I_{min}/I_{max}$ and $P_{max,norm} = I_{max,norm} = 1$. The eye safety regulations [44] and/or the design requirements constrain the level of the radiated average optical power to $P_{avg,norm}$. Through scaling and DC-biasing, the information-carrying signal, x, can be conditioned within the dynamic range of the transmitter front-end in accordance with the average optical power constraint, and the details are presented in Chapter 4.

The single-carrier pulse modulation signals with finite support of intensity values in their PDF can be transmitted with negligible non-linear distortion. However, the O-OFDM signals with large PAPR and infinite support of intensity levels in their PDF experience double-sided signal clipping. It can be described by the non-linear distortion function $F(x) = \Phi(x)$, where $\Phi(x)$ is given as follows:

$$\Phi(x) = \begin{cases} P_{min,norm}, & \text{if } x \leq P_{min,norm}, \\ x, & \text{if } P_{min,norm} < x \leq P_{max,norm}, \\ P_{max,norm}, & \text{if } x > P_{max,norm}. \end{cases} \tag{3.2}$$

Therefore, the normalized non-linear transfer function, $\Psi(s) = F(x) - F(\beta_{DC})$, can be obtained by the use of $F(x) = \Phi(x)$ as follows:

$$\Psi(s) = \begin{cases} \sigma\lambda_{bottom}, & \text{if } s \leq \sigma\lambda_{bottom}, \\ s, & \text{if } \sigma\lambda_{bottom} < s \leq \sigma\lambda_{top}, \\ \sigma\lambda_{top}, & \text{if } s > \sigma\lambda_{top}, \end{cases} \tag{3.3}$$

where λ_{bottom} and λ_{top} are the normalized bottom and top clipping levels.

3.4 Non-linear distortion of Gaussian signals

The O-OFDM time-domain signals follow either a Gaussian or a half-Gaussian distribution due to the central limit theorem (CLT) [136]. Therefore, these signals have a large PAPR. Conditioning them within the limited linear dynamic range of the transmitter can result in non-linear distortion and/or double-sided signal clipping. In the following, an analytical model for the non-linear distortion of the information-carrying subcarriers is presented.

3.4.1 Analysis of generalized non-linear distortion

The undistorted and unclipped continuous-time OFDM symbol, s, follows a zero-mean Gaussian distribution with a variance of $\sigma^2 = P_{s(elec)}$, where $P_{s(elec)}$ is the average electrical symbol power. The DC-biased continuous-time signal to be transmitted through the optical front-end, x, can be expressed as $x = s + \beta_{DC}$. Passing through the LED, the information-carrying signal, x, is subjected to the non-linear transfer function, $F(x)$. The non-linear distortion of the Gaussian and half-Gaussian OFDM symbols in DCO-OFDM and ACO-OFDM, respectively, can be modeled by means of the Bussgang theorem [137] as an attenuation of the data-carrying signal plus a non-Gaussian uncorrelated noise component [110]. Since the received signal is passed through a fast Fourier transform (FFT) at the receiver, the CLT can be applied, and the additive uncorrelated noise from the non-linear distortion is transformed into additive uncorrelated complex-valued Gaussian noise with zero-mean and variance of σ_{clip}^2 [131]. The non-linear distortion parameters such as the gain factor denoting the electrical power attenuation of the OFDM symbol, K, and the variance of the additive non-linear distortion noise, σ_{clip}^2, can be derived for DCO-OFDM and ACO-OFDM by the use of the model for the non-linear transfer function of the optical front-end defined in (3.1) [133].

The non-linear distortion can be modeled by means of the Bussgang theorem for DCO-OFDM and ACO-OFDM, respectively, as follows:

$$\Psi(s) = Ks + w_{\text{clip}} , \tag{3.4}$$

$$\Psi(s) = U(s)2Ks + w_{\text{clip}} . \tag{3.5}$$

Here, $U(\cdot)$ denotes the unit step function which is used to denote the default zero-level clipping of the time-domain signal in ACO-OFDM. The Bussgang theorem states that after the non-linear distortion, the signal is attenuated by a factor, K, and an uncorrelated non-Gaussian noise, w_{clip}, is added. In ACO-OFDM, the amplitude of the received odd subcarriers is reduced by 50% because of the zero-level clipping and the symmetries discussed in [51]. Therefore, the attenuation factor K is multiplied by a factor of 2 in (3.5). In addition, since the received signal is passed through an FFT, the additive non-linear noise, w_{clip}, is transformed into additive zero-mean complex-valued Gaussian noise at the information-carrying subcarriers, preserving its variance of σ_{clip}^2. In the following, K and σ_{clip}^2 are derived for the considered DCO-OFDM and ACO-OFDM systems.

The gain factor of the non-linear distortion denoting the electrical power attenuation of an information-carrying subcarrier in O-OFDM, K, can be expressed in DCO-OFDM and ACO-OFDM from [137] as follows:

$$K = \frac{\text{Cov}\,[s, \Psi(s)]}{\sigma^2} , \tag{3.6}$$

where $\text{Cov}\,[\cdot]$ denotes the covariance operator. Since the OFDM symbol, s, is a zero-mean Gaussian random variable, K can be derived as follows:

$$K = \frac{\text{E}\,[s\Psi(s)]}{\sigma^2} = \frac{1}{\sigma^2} \int_{-\infty}^{+\infty} s\Psi(s)\frac{1}{\sigma}\phi\left(\frac{s}{\sigma}\right) ds , \tag{3.7}$$

where $\text{E}\,[\cdot]$ denotes the expectation operator, and $\phi(\cdot)$ is the PDF of a standard normal distribution with zero-mean and unity variance. Using the structure of the normalized non-linear distortion function, $\Psi(s)$, from (3.1), integration can be performed in a piece-wise fashion as follows:

$$K = \lambda_J \phi(\lambda_J) - \lambda_1 \phi(\lambda_1)$$

$$+ \frac{1}{\sigma^2} \sum_{j=1}^{J-1} \int_{-\infty}^{\sigma \lambda_{j+1}} s \psi_j(s) \frac{1}{\sigma} \phi\left(\frac{s}{\sigma}\right) ds - \frac{1}{\sigma^2} \sum_{j=1}^{J-1} \int_{-\infty}^{\sigma \lambda_j} s \psi_j(s) \frac{1}{\sigma} \phi\left(\frac{s}{\sigma}\right) ds . \tag{3.8}$$

Since $s\psi_j(s)$ are polynomial functions of order $n_j + 1$, the integrals can be expressed as a linear combination of integrals with the following structure:

$$\int_{-\infty}^{\sigma \lambda} s^n \frac{1}{\sigma} \phi\left(\frac{s}{\sigma}\right) ds = \sigma^n \int_{-\infty}^{\lambda} u^n \phi(u) \, du = \sigma^n \mathcal{I}_n . \tag{3.9}$$

The integral, $\mathcal{I}_n$, can be solved using the following recursive relation:

$$\mathcal{I}_n = -\lambda^{n-1} \phi(\lambda) + (n-1)\mathcal{I}_{n-2} , \tag{3.10}$$

where $\mathcal{I}_0 = 1 - Q(\lambda)$ and $\mathcal{I}_1 = -\phi(\lambda)$. Here, $Q(\cdot)$ is the complementary cumulative distribution function (CCDF) of a standard normal distribution with zero-mean and unity variance.

While the analysis of the attenuation factor applies both to DCO-OFDM and ACO-OFDM, the approach for deriving the variance of the non-linear distortion noise differs in the two O-OFDM schemes. Hereafter, the analysis is presented for DCO-OFDM and ACO-OFDM.

In DCO-OFDM, the non-linear distortion noise component, $\mathbf{w}_{\text{clip}}$, can be expressed from (3.4). Thus, the variance of the non-linear noise, σ_{clip}^2, can be derived for the generalized non-linear distortion function, $\Psi(s)$, in DCO-OFDM as follows:

$$\sigma_{\text{clip}}^2 = \mathrm{E}\left[\left(\Psi(s) - Ks\right)^2\right] - \mathrm{E}\left[\Psi(s) - Ks\right]^2 = \mathrm{E}\left[\Psi(s)^2\right] - \mathrm{E}\left[\Psi(s)\right]^2 - K^2 \sigma^2$$

$$= \int_{-\infty}^{+\infty} \Psi(s)^2 \frac{1}{\sigma} \phi\left(\frac{s}{\sigma}\right) ds - \left(\int_{-\infty}^{+\infty} \Psi(s) \frac{1}{\sigma} \phi\left(\frac{s}{\sigma}\right) ds\right)^2 - K^2 \sigma^2 . \tag{3.11}$$

Using (3.1), (3.9), and (3.10), the variance of the non-linear distortion noise, σ_{clip}^2, can be generalized for DCO-OFDM as follows:

$$\sigma_{\text{clip}}^2 = \sigma^2 \lambda_1^2 \left(1 - Q(\lambda_1)\right) + \sigma^2 \lambda_J^2 Q(\lambda_J) - K^2 \sigma^2 + \sum_{j=1}^{J-1} \int_{-\infty}^{\sigma \lambda_{j+1}} \psi_j(s)^2 \frac{1}{\sigma} \phi\left(\frac{s}{\sigma}\right) ds$$

$$- \sum_{j=1}^{J-1} \int_{-\infty}^{\sigma \lambda_j} \psi_j(s)^2 \frac{1}{\sigma} \phi\left(\frac{s}{\sigma}\right) ds - \left(\sigma \lambda_1 \left(1 - Q(\lambda_1)\right) + \sigma \lambda_J Q(\lambda_J)\right) \tag{3.12}$$

$$+ \sum_{j=1}^{J-1} \int_{-\infty}^{\sigma \lambda_{j+1}} \psi_j(s) \frac{1}{\sigma} \phi\left(\frac{s}{\sigma}\right) ds - \sum_{j=1}^{J-1} \int_{-\infty}^{\sigma \lambda_j} \psi_j(s) \frac{1}{\sigma} \phi\left(\frac{s}{\sigma}\right) ds\Bigg)^2 .$$

In order to derive the variance of the non-linear noise component in ACO-OFDM, the half-Gaussian distribution of the undistorted ACO-OFDM symbol, s, is unfolded by mirroring of the positive signal samples around the origin as illustrated in Fig. 3.7 in Section 3.4.2. The resulting unfolded symbol, $\hat{s}$, follows a zero-mean real-valued Gaussian distribution with a variance of $\sigma^2/2$. The corresponding unfolded non-linear distortion function, $\hat{\Psi}(\hat{s})$, is symmetric with respect to the origin, and it can be written as follows:

$$\hat{\Psi}(\hat{s}) = \begin{cases} \Psi\left(s/\sqrt{2}\right), & \text{if } \hat{s} \geq 0, \\ \Psi\left(-s/\sqrt{2}\right), & \text{if } \hat{s} < 0. \end{cases} \tag{3.13}$$

The unfolded signal, $\hat{s}$, has a bias of $-\sigma\lambda_1/\sqrt{2}$ on the negative samples and a bias of $\sigma\lambda_1/\sqrt{2}$ on the positive ones. Since these biases are applied to the subcarriers with indices $\{0, N/2\}$ in the ACO-OFDM frame after the FFT block at the receiver, they are irrelevant to the non-linear noise variance on the data-carrying subcarriers. Therefore, these biases are removed as follows:

$$\bar{s} = \begin{cases} \hat{s} - \sigma\lambda_1/\sqrt{2}, & \text{if } \hat{s} \geq 0, \\ \hat{s} + \sigma\lambda_1/\sqrt{2}, & \text{if } \hat{s} < 0. \end{cases} \tag{3.14}$$

As a result, the variance of the non-linear noise component, σ_{clip}^2, can be derived for the generalized non-linear distortion function, $\Psi(s)$, in ACO-OFDM from (3.5) as follows:

$$\begin{aligned} \sigma_{\text{clip}}^2 &= \mathrm{E}\left[\left(\hat{\Psi}(\bar{s}) - 2K\hat{s}\right)^2\right] - \mathrm{E}\left[\hat{\Psi}(\bar{s}) - 2K\hat{s}\right]^2 \\ &= \mathrm{E}\left[\hat{\Psi}(\bar{s})^2\right] - 4K\mathrm{E}\left[\hat{\Psi}(\bar{s})\hat{s}\right] + 4K^2\mathrm{E}\left[\hat{s}^2\right] \\ &\quad - \mathrm{E}\left[\hat{\Psi}(\bar{s})\right]^2 + 4K\mathrm{E}\left[\hat{\Psi}(\bar{s})\right]\mathrm{E}\left[\hat{s}\right] - 4K^2\mathrm{E}\left[\hat{s}\right]^2. \end{aligned} \tag{3.15}$$

Given that $\mathrm{E}\left[\hat{\Psi}(\bar{s})^2\right] = \mathrm{E}\left[\Psi(s - \sigma\lambda_1)^2\right]$, $\mathrm{E}\left[\hat{\Psi}(\bar{s})\hat{s}\right] = K\sigma^2$, $\mathrm{E}\left[\hat{\Psi}(\bar{s})\right] = 0$, and $\mathrm{E}\left[\hat{s}\right] = 0$, σ_{clip}^2 can be expressed in ACO-OFDM as follows:

$$\sigma_{\text{clip}}^2 = \mathrm{E}\left[\Psi(s - \sigma\lambda_1)^2\right] - 2K^2\sigma^2 = \int_{-\infty}^{+\infty}\Psi(s - \sigma\lambda_1)^2\frac{1}{\sigma}\phi\left(\frac{s}{\sigma}\right)ds - 2K^2\sigma^2, \tag{3.16}$$

where the debiased normalized piecewise polynomial function, $\Psi(s - \sigma\lambda_1)$, on the half-Gaussian ACO-OFDM symbol, s, is defined as follows:

$$\Psi(s - \sigma\lambda_1) = \begin{cases} 0, & \text{if } s \leq \sigma\lambda_1, \\ \psi_1(s - \sigma\lambda_1), & \text{if } \sigma\lambda_1 < s \leq \sigma\lambda_2, \\ \vdots & \\ \psi_{L-1}(s - \sigma\lambda_1), & \text{if } \sigma\lambda_{L-1} < s \leq \sigma\lambda_L, \\ \sigma(\lambda_L - \lambda_1), & \text{if } s > \sigma\lambda_L. \end{cases} \tag{3.17}$$

Using (3.17), (3.9), and (3.10), the variance of the non-linear distortion noise, σ_{clip}^2, can be generalized for ACO-OFDM as follows:

$$\sigma_{\text{clip}}^2 = \sigma^2(\lambda_J - \lambda_1)^2 Q(\lambda_J) - 2K^2\sigma^2 + \sum_{j=1}^{J-1} \int_{-\infty}^{\sigma\lambda_{j+1}} \psi_j(s - \sigma\lambda_1)^2 \frac{1}{\sigma}\phi\left(\frac{s}{\sigma}\right) ds$$

$$ - \sum_{j=1}^{J-1} \int_{-\infty}^{\sigma\lambda_j} \psi_j(s - \sigma\lambda_1)^2 \frac{1}{\sigma}\phi\left(\frac{s}{\sigma}\right) ds .$$

(3.18)

The integrals in (3.12) and (3.18) can be solved using the structure from (3.9) and (3.10). In addition, the expressions in (3.8), (3.12), and (3.18) do not exhibit a dominant term, and all of the addends are required to accurately model the non-linear distortion.

3.4.2 Analysis of double-sided signal clipping distortion

After pre-distortion, the OFDM time-domain symbol, s, is subjected to double-sided clipping at normalized bottom and top clipping levels, λ_{bottom} and λ_{top}. This non-linear transfer function of the optical front-end is given in (3.3). This linearized and clipped normalized transfer function can be used in (3.4) and (3.5) to obtain the clipping noise component, w_{clip}. In ACO-OFDM, w_{clip} is non-negative, and it has a unipolar distribution, whereas in DCO-OFDM the clipping noise is bipolar. Passing through the FFT at the receiver, the clipping noise, w_{clip}, is transformed into additive zero-mean complex-valued Gaussian noise at the information-carrying subcarriers, preserving its variance of σ_{clip}^2. In the following, K and σ_{clip}^2 are derived for the case of double-sided signal clipping after pre-distortion in the considered DCO-OFDM and ACO-OFDM systems [131].

In DCO-OFDM and ACO-OFDM, the attenuation factor of the information-carrying subcarriers, K, can be expressed from (3.7) and (3.8) using (3.9) and (3.10) as follows:

$$K = \lambda_{\text{top}}\phi(\lambda_{\text{top}}) - \lambda_{\text{bottom}}\phi(\lambda_{\text{bottom}})$$

$$ + \frac{1}{\sigma^2}\int_{-\infty}^{\sigma\lambda_{\text{top}}} s^2 \frac{1}{\sigma}\phi\left(\frac{s}{\sigma}\right) ds - \frac{1}{\sigma^2}\int_{-\infty}^{\sigma\lambda_{\text{bottom}}} s^2 \frac{1}{\sigma}\phi\left(\frac{s}{\sigma}\right) ds$$

(3.19)

$$ = Q(\lambda_{\text{bottom}}) - Q(\lambda_{\text{top}}) .$$

Note that in ACO-OFDM and DCO-OFDM, since s is real, K is a real-valued function. The cattenuation factor essentially represents the likelihood of samples not being clipped. In addition, (3.19) shows that K is independent of the modulation scheme, M-QAM, and the FFT size, N, under the assumption of a Gaussian distribution of the unclipped signal which holds true for practical O-OFDM systems with $N > 64$. The attenuation factor as a function of the normalized bottom and top clipping levels is illustrated in Figs. 3.4 and 3.5 for DCO-OFDM and ACO-OFDM, respectively. The signal clipping in DCO-OFDM exhibits a symmetric attenuation profile for clipping levels located symmetrically around the average optical power level. As suggested

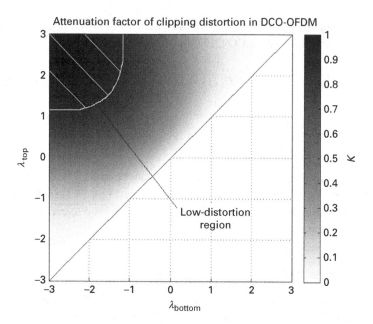

Figure 3.4 Attenuation factor of the clipping noise as a function of the normalized clipping levels in DCO-OFDM [132]. ©2011 IEEE

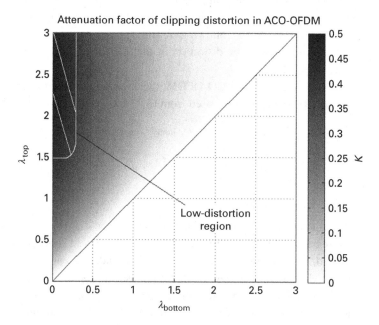

Figure 3.5 Attenuation factor of the clipping noise as a function of the normalized clipping levels in ACO-OFDM [132]. ©2011 IEEE

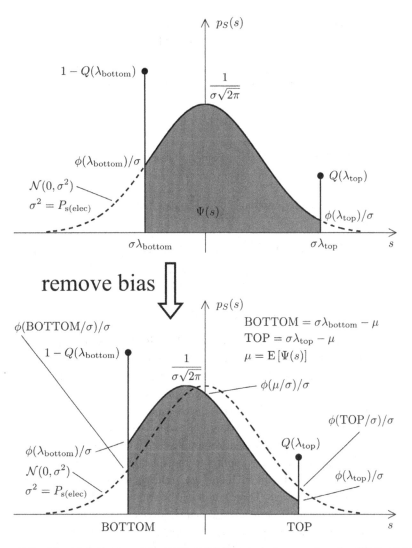

Figure 3.6 Double-sided clipping of the DCO-OFDM symbol after pre-distortion [131]. ©2012 IEEE

in [51, 138], the attenuation factor for ACO-OFDM approaches 0.5 when the least signal clipping is present. Furthermore, the ACO-OFDM symbol experiences larger attenuation for downside clipping as compared to upside clipping.

The distorted DCO-OFDM symbol, $\Psi(s)$, is illustrated in Fig. 3.6. It follows a close to Gaussian distribution with zero-mean and variance of $P_{s(elec)}$, when the least signal clipping is present. Clipping the OFDM symbol at normalized bottom and top clipping levels, λ_{bottom} and λ_{top}, in an asymmetric fashion, results in a non-zero-mean signal, $\Psi(s)$, which has a bias of $E[\Psi(s)]$, where $E[\cdot]$ denotes the expectation operator. The bias of $E[\Psi(s)]$ can be removed as shown in Fig. 3.6. In DCO-OFDM, the clipping noise variance, σ_{clip}^2, can be derived for the normalized non-linear distortion function of double-sided signal clipping from (3.3) by the use of (3.11) and (3.12). Here, the integrals for $E\left[\Psi(s)^2\right]$ and $E\left[\Psi(s)\right]$ can be solved by means of (3.9) and (3.10) for (3.3) as follows:

$$
\begin{aligned}
E\left[\Psi(s)^2\right] = P_{s(\text{elec})}\Big(&K + \phi(\lambda_{\text{bottom}})\lambda_{\text{bottom}} - \phi(\lambda_{\text{top}})\lambda_{\text{top}} \\
&+ \left(1 - Q(\lambda_{\text{bottom}})\right)\lambda_{\text{bottom}}^2 + Q(\lambda_{\text{top}})\lambda_{\text{top}}^2\Big),
\end{aligned}
\tag{3.20}
$$

$$
E\left[\Psi(s)\right] = \sqrt{P_{s(\text{elec})}}\Big(\phi(\lambda_{\text{bottom}}) - \phi(\lambda_{\text{top}}) + \left(1 - Q(\lambda_{\text{bottom}})\right)\lambda_{\text{bottom}} + Q(\lambda_{\text{top}})\lambda_{\text{top}}\Big).
\tag{3.21}
$$

As a result, the clipping noise variance, σ_{clip}^2, in DCO-OFDM can be expressed as follows:

$$
\begin{aligned}
\sigma_{\text{clip}}^2 = P_{s(\text{elec})}\Big(&K - K^2 + (1 - Q(\lambda_{\text{bottom}}))\lambda_{\text{bottom}}^2 + Q(\lambda_{\text{top}})\lambda_{\text{top}}^2 \\
&- \left(\phi(\lambda_{\text{bottom}}) - \phi(\lambda_{\text{top}}) + (1 - Q(\lambda_{\text{bottom}}))\lambda_{\text{bottom}} + Q(\lambda_{\text{top}})\lambda_{\text{top}}\right)^2 \\
&+ \phi(\lambda_{\text{bottom}})\lambda_{\text{bottom}} - \phi(\lambda_{\text{top}})\lambda_{\text{top}}\Big).
\end{aligned}
\tag{3.22}
$$

The distorted ACO-OFDM symbol, $\Psi(s)$, is illustrated in Fig. 3.7. Here, the symmetries discussed in [51] allow for the unfolding of the truncated half-Gaussian distribution of $\Psi(s)$, the mirroring of the clipping levels around the origin, and the redistribution of the signal samples. The resulting signal is symmetric with respect to the origin, and it follows a close to Gaussian distribution with zero-mean and variance of $P_{s(\text{elec})}/2$, when the least signal clipping is present. However, the unfolded signal has a bias of $-\sigma\lambda_{\text{bottom}}/\sqrt{2}$ on the negative samples and a bias of $\sigma\lambda_{\text{bottom}}/\sqrt{2}$ on the positive ones. Since these biases are applied to the subcarriers with indices $\{0, N/2\}$ in the ACO-OFDM frame after the FFT, they are irrelevant to the clipping noise variance on the data-carrying subcarriers. These biases can be removed as shown in Fig. 3.7. In ACO-OFDM, the clipping noise variance, σ_{clip}^2, can be derived for the normalized non-linear distortion function of double-sided signal clipping from (3.3) by the use of (3.16) and (3.18). Here, the integral for $E\left[\Psi(s - \sigma\lambda_{\text{bottom}})^2\right]$ can be solved by means of (3.9) and (3.10) for (3.3) as follows:

$$
\begin{aligned}
E\left[\Psi(s - \sigma\lambda_{\text{bottom}})^2\right] = P_{s(\text{elec})}\Big(&K\left(\lambda_{\text{bottom}}^2 + 1\right) + Q(\lambda_{\text{top}})\left(\lambda_{\text{top}} - \lambda_{\text{bottom}}\right)^2 \\
&+ 2\phi(\lambda_{\text{top}})\lambda_{\text{bottom}} - \phi(\lambda_{\text{top}})\lambda_{\text{top}} - \phi(\lambda_{\text{bottom}})\lambda_{\text{bottom}}\Big).
\end{aligned}
\tag{3.23}
$$

As a result, the clipping noise variance, σ_{clip}^2, in ACO-OFDM can be expressed as follows:

$$
\begin{aligned}
\sigma_{\text{clip}}^2 = P_{s(\text{elec})}\Big(&K\left(\lambda_{\text{bottom}}^2 + 1\right) - 2K^2 - \lambda_{\text{bottom}}\left(\phi(\lambda_{\text{bottom}}) - \phi(\lambda_{\text{top}})\right) \\
&- \phi(\lambda_{\text{top}})\left(\lambda_{\text{top}} - \lambda_{\text{bottom}}\right) + Q(\lambda_{\text{top}})\left(\lambda_{\text{top}} - \lambda_{\text{bottom}}\right)^2\Big).
\end{aligned}
\tag{3.24}
$$

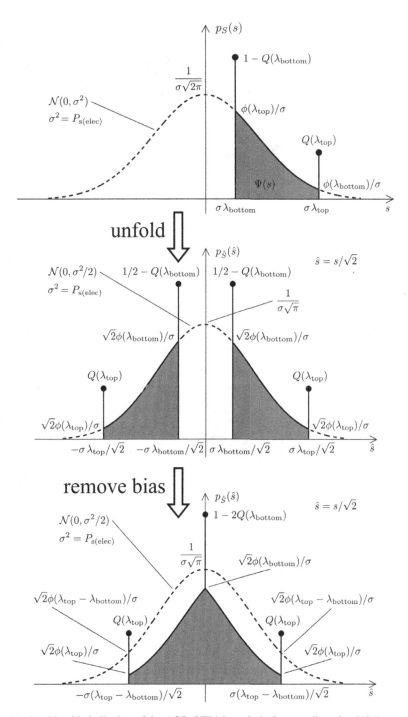

Figure 3.7 Double-sided clipping of the ACO-OFDM symbol after pre-distortion [131]. ©2012 IEEE

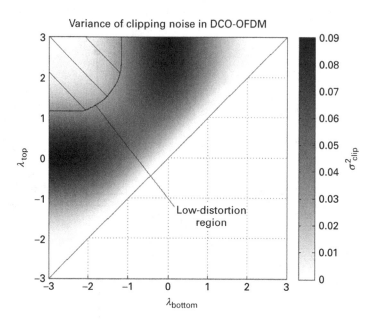

Figure 3.8 Variance of the clipping noise as a function of the normalized clipping levels in DCO-OFDM [132]. Here, a scaled symbol power of $P_{s(elec)}/G_B = 1$ is assumed. ©2011 IEEE

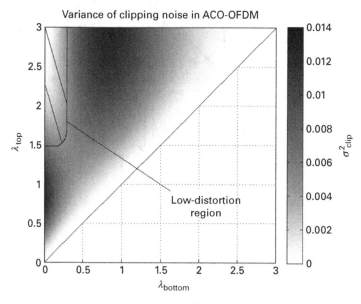

Figure 3.9 Variance of the clipping noise as a function of the normalized clipping levels in ACO-OFDM [132]. Here, a scaled symbol power of $P_{s(elec)}/G_B = 1$ is assumed. ©2011 IEEE

Note that the clipping noise variance, σ_{clip}^2, is also shown to be independent of the modulation order, M, and the FFT size, N, as long as the PDF of the unclipped signal is close to Gaussian. It only depends on the normalized bottom and top clipping levels. This is illustrated in Figs. 3.8 and 3.9 for DCO-OFDM and ACO-OFDM, respectively. As expected, σ_{clip}^2 approaches zero for the least signal clipping. However, since K and

σ_{clip}^2 are coupled according to (3.4) and (3.5), as K approaches zero, σ_{clip}^2 approaches zero as well. Here, the large symbol distortion is defined for a small K and a large σ_{clip}^2. As expected, it is shown in Figs. 3.4 and 3.8 that the symbol distortion in DCO-OFDM can be minimized when symmetric normalized clipping levels are chosen. In addition, Figs. 3.5 and 3.9 suggest that downside clipping introduces a larger ACO-OFDM symbol distortion than upside clipping. Since K and σ_{clip}^2 are independent of M, they remain constant across the modulation orders for a particular choice of normalized bottom and top clipping levels. Nonetheless, the BER performance of the system depends on the granularity of the constellation, where higher-order modulation is more vulnerable to Gaussian noise because of the shrinking of the decision regions. Therefore, the non-linear clipping distortion more significantly impacts the BER performance of higher-order modulation, M-QAM.

3.5   Summary

In this chapter, an analytical model for the non-linear transfer function of the optical front-end was presented. A piecewise polynomial model has been proposed as a flexible and accurate generalized representation of the non-linear transfer of the information-carrying current. The non-linear distortion of the large PAPR O-OFDM signals conditioned within the limited dynamic range of the transmitter has been analyzed in detail. The non-linear distortion has been translated by means of the Bussgang theorem and the CLT as an attenuation of the information-carrying subcarriers at the receiver, plus zero-mean Gaussian uncorrelated noise. The non-linear distortion parameters, such as the attenuation factor and the clipping noise variance, have been derived in closed form for the generalized piecewise polynomial transfer function of the transmitter. Practical FFT sizes greater than 64, which ensure Gaussianity of the time-domain O-OFDM signal, have been proven not to affect the attenuation factor and the non-linear noise variance. In addition, the non-linear distortion parameters have been shown to be independent of the modulation scheme of the OFDM subcarriers. However, higher-order M-QAM is expected to experience a larger BER degradation for a given signal biasing setup, i.e. signal scaling and DC bias.

The inverse of the piecewise polynomial transfer function has been shown to be applicable for signal pre-distortion in order to maximize the linear dynamic range. As a result, the non-linear distortion in O-OFDM has been reduced to double-sided signal clipping. The attenuation factor of the clipping distortion and the clipping noise variance in O-OFDM have been derived in closed form also for the case of double-sided signal clipping after pre-distortion. The non-linear distortion of the DCO-OFDM signal has been shown to be minimized for a symmetric clipping setup, while in ACO-OFDM the bottom-level clipping introduced larger signal distortion as compared to top-level clipping.

4 Digital modulation schemes

4.1 Introduction

The data transmission in optical wireless communication (OWC) with incoherent light sources is realized through intensity modulation and direct detection (IM/DD). For this purpose, the transmitted signal needs to be real-valued and non-negative. In practice, this is achieved by single-carrier modulation techniques, such as multi-level pulse position modulation (M-PPM) and multi-level pulse amplitude modulation (M-PAM), and through multi-carrier modulation such as multi-level quadrature amplitude modulation (M-QAM) optical orthogonal frequency division multiplexing (O-OFDM). Conventionally, the average optical power is defined as the first moment of the transmitted signal, while the average electrical power is defined as the second moment of the transmitted signal. In practice, the dynamic range can be linearized through pre-distortion only between levels of minimum and maximum radiated optical power. In addition, eye safety regulations [44] and/or design requirements also impose an average optical power constraint. Because of these constraints, there is a fixed relation between average electrical power and the average optical power of the single-carrier and multi-carrier signals which varies with the change in the biasing setup, i.e. a combination of direct current (DC) bias and signal variance. The optical-to-electrical (O/E) conversion of the optical signals is investigated in this chapter. In this study, the electrical energy consumption of the OWC system is considered, and therefore the average electrical power carries the information [67]. Here, the received electrical signal-to-noise ratio (SNR) is presented as a function of the channel equalization penalty, the DC-bias penalty, and the non-linear distortion parameters. The analytical framework is verified by means of a Monte Carlo bit-error ratio (BER) simulation [131, 133, 134].

4.2 Optical signals

As opposed to a radio frequency (RF) system, where the data-carrying signal modulates the complex-valued bipolar electric field radiated by an antenna, in an OWC system the signal modulates the intensity of the optical emitter, and therefore it needs to be real-valued and unipolar non-negative. Since light emitting diodes (LEDs) are incoherent light sources, it is difficult to collect signal power in a single electromagnetic mode and to provide a stable carrier in an indoor OWC scenario. Therefore, it is infeasible to construct an efficient coherent receiver, such as the superheterodyne receiver commonly used in RF transmission. Practical low-cost optical carrier modulation for OWC is usually achieved through IM/DD in an incoherent fashion. The information is encoded in the envelope of the transmitted signal, and there is no phase information.

Single-carrier pulse modulation is a suitable candidate for data transmission in OWC. The information can be encoded in the duration of the pulse, such as pulse width modulation (PWM) or pulse interval modulation (PIM) [4, 139]. In addition, the information can be encoded in the position of the pulse, such as M-PPM [20, 21]. Alternatively, the information can be encoded in the amplitude of the pulse, such as M-PAM [45]. In this study, M-PPM and M-PAM are considered as benchmark schemes. Examples for binary and multi-level realizations of these single-carrier modulation techniques are given in Figs. 4.1 and 4.2, respectively. The limited linear dynamic range of the transmitter between positive levels of minimum and maximum forward currents, i.e. I_{min} and I_{max}, is considered, resulting in an additional DC-bias requirement in the M-PAM. Here, low-frequency distortion effects from DC wander in the electrical components, as well as from the flickering of background light sources, are known to degrade the SNR. By encoding the information in the position of the pulse within the transmitted symbol, M-PPM has been considered for OWC in [140] because of its ability to provide robust low-complexity low-rate transmission at very low SNR in the flat fading channel. Furthermore, trellis coding of the M-PPM signal is shown to increase the OWC system robustness and also spectral efficiency in a dispersive channel [141]. Here, lower SNR requirement for a target BER with a lower spectral efficiency is

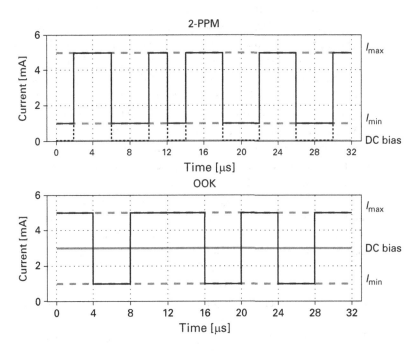

Figure 4.1 Single-carrier binary optical transmission

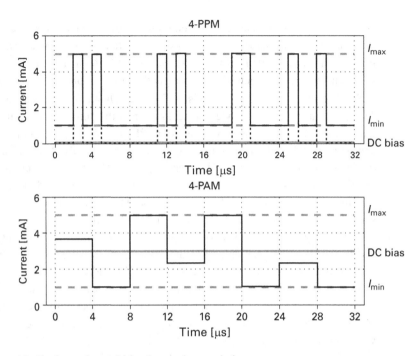

Figure 4.2 Single-carrier multi-level optical transmission

achieved by increasing the modulation order, M, where 2-PPM has the highest SNR requirement and spectral efficiency. Showing the same SNR requirement as 2-PPM, a binary on–off keying (OOK) transmission doubles the spectral efficiency, and therefore it has been the preferred candidate for the pursuit of higher data rates in an OWC system [3, 5–7]. In addition, if the link budget accommodates higher SNR, the spectral efficiency of the single-carrier system can be increased by encoding the information in the amplitude of the pulse, i.e. amplitude of the transmitted symbol, such as multi-level M-PAM [45, 142]. The benefits of M-PPM and M-PAM are united in the hybrid system, i.e. multi-level pulse amplitude and position modulation (M-PAPM) [143].

At low data rates, the 3-dB coherence bandwidth of the dispersive optical wireless channel is larger than the 3-dB bandwidth of the information-carrying pulse, and therefore the dispersive channel can be considered as flat fading over the pulse bandwidth. In single-carrier optical wireless systems, the challenge arises in the attempt to push high data rates through the dispersive channel [25, 75, 140, 144, 145]. At high data rates, where the pulse bandwidth exceeds the channel coherence bandwidth, the root-mean-squared (RMS) delay spread of the channel impulse response exceeds the pulse duration. Therefore, such techniques experience severe inter-symbol interference (ISI), limiting their throughput. In general, a guard interval can be added to reduce the overlap between the information-carrying pulses [146] at the expense of decreased system spectral efficiency. In order to compensate for the channel effect, the optimum receiver employs maximum likelihood sequence detection (MLSD) [21, 45]. Here, the MLSD algorithm chooses the sequence of symbols that maximizes the likelihood of the received symbols with the knowledge of the channel taps. Even though the Viterbi algorithm can be used for MLSD to reduce the computational effort, the complexity of MLSD still increases exponentially with the number of channel taps. Therefore, in practical system implementations, suboptimum receivers employ equalization techniques with feasible complexity. These include the linear feed-forward equalizer (FFE) or the non-linear decision feedback equalizer (DFE) with zero forcing (ZF) or minimum mean-squared error (MMSE) criteria [6, 45]. The superior BER performance at a lower SNR requirement of DFE comes at significantly increased complexity compared to FFE [45].

Multi-carrier modulation such as M-QAM O-OFDM has inherent robustness to ISI because the symbol duration is significantly longer than the RMS delay spread of the optical wireless channel. In addition, O-OFDM is immune to low-frequency distortion from background lighting and DC wander. As a result, M-QAM O-OFDM promises to deliver very high data rates [9, 47, 67]. Because of the common use of a cyclic prefix (CP), the channel frequency response can be considered as flat fading over the subcarrier bandwidth [107, 108]. Thus, single-tap linear FFE with low complexity paired with bit and power loading can be used to minimize the channel effect [49, 50]. In O-OFDM, the time-domain signal envelope is utilized to modulate the intensity of the LED. For this purpose, the signal needs to be real-valued and non-negative. A real-valued signal is obtained when Hermitian symmetry is imposed on the OFDM subcarriers. One approach to obtain a non-negative signal, known as direct-current-biased O-OFDM (DCO-OFDM), is the addition of a DC bias to the OFDM symbol, s [147, 148]. A closely related technique, the discrete multi-tone (DMT), has been employed for digital subscriber line (DSL) data transmission [149]. In addition to these two commonly used O-OFDM Gaussian signals, there has been significant effort to design O-OFDM unipolar half-Gaussian signals for applications with very low radiated optical power. First, Armstrong *et al.* proposed a technique known as asymmetrically clipped O-OFDM (ACO-OFDM) [51, 107, 138]. By enabling the odd subcarriers for data transmission and setting the even ones to zero, the negative part of the time-domain signal can be clipped at the transmitter. The clipping noise falls on the even subcarriers, and therefore there is no distortion on the information-carrying odd subcarriers at the receiver. In comparison to DCO-OFDM, ACO-OFDM is expected to achieve better spectral efficiency in the low-SNR region at the expense of a 50% reduction in spectral efficiency at high SNRs. An example of O-OFDM transmission is illustrated in Fig. 4.3 for a limited linear dynamic range of the transmitter, resulting in non-linear signal clipping distortion [150, 151]. A modulation technique closely related to ACO-OFDM with a similar electrical SNR requirement and spectral efficiency has been presented in [52] and named as pulse amplitude modulation discrete multi-tone (PAM-DMT). The difference between the two lie within two steps of the modulation procedures. In PAM-DMT, only the imaginary part of the information-carrying odd subcarriers is modulated in an M-PAM fashion, while the real parts are set to zero, and there is no Hermitian symmetry imposed on the OFDM frame. Another unipolar modulation approach with a similar SNR requirement and spectral efficiency is known as flip-OFDM [53]. By exploiting the symmetry within the time-domain signal samples [51], computational complexity can be reduced by 50% by halving the number of fast Fourier transform (FFT) operations.

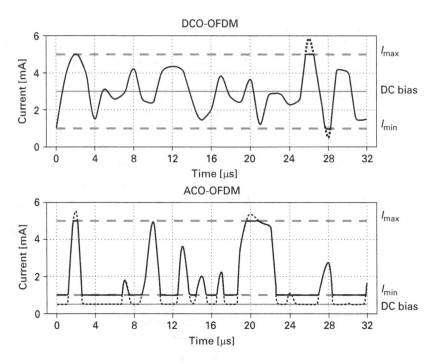

Figure 4.3 Multi-carrier multi-level optical transmission

In addition, there have been attempts to improve the electrical power efficiency of the O-OFDM unipolar half-Gaussian signals [54, 152]. Again by exploiting the symmetry within the time-domain signal samples [51], the signal demodulation procedure of ACO-OFDM can be improved by selecting the pairs of corresponding low and high sample values. Then, the high sample value is identified by means of a maximum-level detector, while the low sample value is set to zero. This results in an effective reduction of the additive white Gaussian noise (AWGN), corresponding to a reduction in the electrical SNR requirement of approximately 1 dB [152]. However, such a procedure only works well in a flat fading channel because the ISI in the dispersive channel significantly degrades the accuracy of the maximum-level decision. This detector structure can be combined with the flip-OFDM structure to also achieve a 50% reduction in computational complexity. The combined modulation scheme is named as unipolar OFDM (U-OFDM) and presented in [54]. Since the improvements are in addition to optimum signal scaling and DC-biasing to counter non-linear signal distortion, hereafter only the general ACO-OFDM scheme is considered for modeling and comparison purposes.

Since the M-PPM and M-PAM signals have a probability density function (PDF) with finite support, and because of their respective demodulation procedures, they can fit the minimum, average, and maximum optical power constraints of the transmitter without effective non-linear clipping distortion. The O-OFDM systems use the inverse FFT (IFFT) as a multiplexing technique at the transmitter. Therefore, for a large number of subcarriers the non-distorted time-domain signals in DCO-OFDM and ACO-OFDM closely follow Gaussian and half-Gaussian distributions, respectively, according to the central limit theorem (CLT) [136]. A total subcarrier number as small as 64 is sufficient to ensure Gaussianity [110]. Thus, signal scaling and DC-biasing in DCO-OFDM and ACO-OFDM result in non-linear signal distortion, which can be modeled by means of the Bussgang theorem [137] as an attenuation of the data-carrying signal plus a non-Gaussian uncorrelated clipping noise component [110, 142, 153]. At the receiver, the FFT is used for demultiplexing. Therefore, the CLT can be applied again, and the clipping noise component can be modeled as a Gaussian process. This methodology is used in [110], where the non-linear transfer effects due to the short dynamic range of high-power amplifiers (HPA) in OFDM-based RF systems are studied. Symmetric signal clipping due to the large peak-to-average-power ratio (PAPR) in RF OFDM is studied in [154, 155]. Equivalently, symmetric signal clipping in O-OFDM, e.g. DMT for DSL transmission, is investigated in [142, 153, 156, 157]. An optimal DC bias of a symmetrically clipped signal can be inferred from [157, 158]. It is shown in [157] that iterative decoding with clipping noise estimation and subtraction can reduce the BER for increased computational complexity. In this study, a comprehensive analysis is presented for a generalized non-linear transfer function and for a double-sided signal clipping at independent bottom and top levels after pre-distortion. In addition, the analysis is employed in the formulation of optimum signal scaling and DC-biasing to minimize the required electrical SNR per bit for a target BER in a practical dynamic range of the transmitter.

OOK, essentially 2-PAM, and M-PPM have been compared in terms of the electrical and the optical power requirement in a dispersive channel with equalization in [21]. An increasing power requirement is demonstrated with the increase of the RMS channel delay spread or, equivalently, data rate. In a later study [159], M-PPM, M-PAM, and multi-carrier M-QAM transmission, similar to M-QAM DCO-OFDM, have been compared assuming a flat fading channel in terms of the optical power requirement and spectral efficiency. However, a positive infinite linear dynamic range of the transmitter is considered which is not generally achievable in practice. In addition, the non-negative M-QAM signal is scaled down to accommodate the large PAPR, resulting in an increased optical power requirement. Recently, a similar comparison has been reported in [138] for the multi-carrier transmission schemes ACO-OFDM and DCO-OFDM with a tolerable clipping distortion. In this book, single-carrier and multi-carrier transmission schemes are compared in terms of the spectral efficiency and electrical SNR requirement in a dispersive realistic optical wireless channel with optimum signal scaling and DC-biasing for a practical dynamic range of the transmitter, where the DC-bias power is excluded or included in the calculation of the SNR.

4.3   Single-carrier modulation

In single-carrier pulse modulation, the information can be encoded either in the position of the pulse, i.e. M-PPM, or in the amplitude of the pulse, i.e. M-PAM. In the following, the details of the O/E conversion of these signals are presented. The linear dynamic range of the transmitter between positive levels of minimum and maximum radiated optical power is considered, and the signals are conditioned within the range through signal scaling and DC-biasing. The electrical SNR at the receiver and the resulting BER performance are derived and verified by means of a Monte Carlo simulation.

4.3.1   Pulse position modulation: M-PPM

The block diagram of single-carrier transmission with pulse modulation is presented in Fig. 4.4. In M-PPM, $\log_2(M)$ equiprobable input bits form a time-domain symbol. It is a sequence of M chips represented as a vector, $\mathbf{c}$, where one chip has a current level of $\sqrt{MP_{s(elec)}}$, and the other $M - 1$ chips are set to zero. Here, $P_{s(elec)}$ is the average electrical power of the M-PPM symbol, and it is related to the average electrical bit power, $P_{b(elec)}$, as follows: $P_{s(elec)} = P_{b(elec)} \log_2(M)$. The respective energy per symbol, $E_{s(elec)}$, and energy per bit, $E_{b(elec)}$, are obtained from the symbol rate, R_s, as follows: $E_{s(elec)} = P_{s(elec)}/R_s$ and $E_{b(elec)} = P_{b(elec)}/R_s$. Here, the M chips with duration of T_s/M fit within a time period of T_s. Therefore, the M-PPM symbol with a bandwidth of $B = M/T_s$ has a duration of T_s for a symbol rate of $R_s = B/M$, and it is grouped

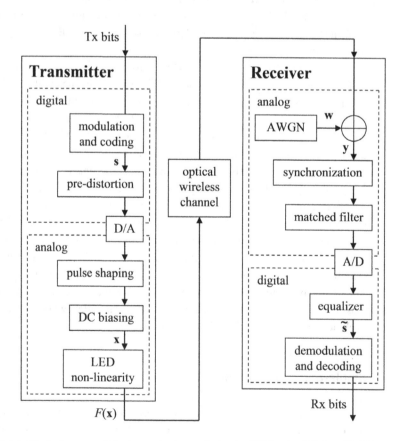

Figure 4.4 Block diagram of single-carrier transmission in OWC using PPM and PAM

in the train of L symbols, $\mathbf{s}_l$, where l, $l = 0, 1, \ldots, L - 1$, is the symbol index. The spectral efficiency of a modulation scheme is determined by the number of bits that can be transmitted per symbol duration and per symbol bandwidth, i.e. SE $= R_b/B$, where $R_b = \log_2(M)R_s$ is the bit rate. Thus, the spectral efficiency of M-PPM is $\log_2(M)/M$ bits/s/Hz [21, 45]. The train of symbols, $\mathbf{s}_l$, is pre-distorted by the inverse of the non-linear transfer function of the LED transmitter, and it is scaled by a factor, α, in order to fit within the front-end optical power constraints. Next, the signal is passed through a digital-to-analog converter (DAC) to transform the train of digital chips into a train of continuous-time pulses. A pulse shaping filter with a real-valued impulse response of $v(t)$ is applied to obtain a band-limited signal. In M-PPM, because of the fact that the information-carrying pulse has an optical power level greater than $P_{\text{min,norm}}$, zero bias is required. As a result, the transmitted signal vector, $F(\mathbf{x}_l)$, has a length of $Z_\mathbf{x} = LM$, and it can be expressed as follows:

$$F(\mathbf{x}_l) = \alpha \mathbf{s}_l \,, \tag{4.1}$$

where

$$\alpha = \min\left(P_{\text{max,norm}}, MP_{\text{avg,norm}}\right)/\sqrt{MP_{\text{s(elec)}}} \,. \tag{4.2}$$

The transmitter front-end constrains $P_{\text{min,norm}}$ and $P_{\text{max,norm}}$, whereas $P_{\text{avg,norm}}$ is independently imposed by the eye safety regulations and/or the design requirements. In M-PPM, when $MP_{\text{avg,norm}} < P_{\text{max,norm}}$, the resulting pulse level, $MP_{\text{avg,norm}}$, should be set greater than $P_{\text{min,norm}}$ in order to turn on the transmitter. The average optical power of the signal is denoted as $\mathrm{E}\,[F(\mathbf{x}_l)]$ where $\mathrm{E}\,[\cdot]$ denotes the expectation operator. In general, constraining the average optical power level to $\mathrm{E}\,[F(\mathbf{x}_l)] \leq P_{\text{avg,norm}}$ results in suboptimum BER performance of the OWC systems. The best BER performance is obtained when this constraint is relaxed, i.e. when $\mathrm{E}\,[F(\mathbf{x}_l)]$ is allowed to assume any level in the dynamic range between $P_{\text{min,norm}}$ and $P_{\text{max,norm}}$.

In order to relate the average optical symbol power, $P_{\text{s(opt)}}$, to the electrical symbol power, $P_{\text{s(elec)}}$, the signal is subjected to O/E conversion defined as follows:

$$P_{\text{s(elec)}} = \frac{\mathrm{E}\left[F(\mathbf{x}_l)^2\right]}{\mathrm{E}\left[F(\mathbf{x}_l)\right]^2} P_{\text{s(opt)}}^2 \,, \tag{4.3}$$

where $\mathrm{E}\,[F(\mathbf{x}_l)]^2 = \min(P_{\text{max,norm}}/M, P_{\text{avg,norm}})^2$ in M-PPM. In addition, the second moment is given as: $\mathrm{E}\left[F(\mathbf{x}_l)^2\right] = \min(P_{\text{max,norm}}, MP_{\text{avg,norm}})^2/M$. Since the time-domain signal in M-PPM has a PDF with finite support, and taking into account its demodulation procedure elaborated below, it can be fitted within $P_{\text{min,norm}}$ and $P_{\text{max,norm}}$ without effective non-linear clipping distortion. Thus, the following holds for its average optical signal power: $\mathrm{E}\,[F(\mathbf{x}_l)] = P_{\text{s(opt)}} = \min(P_{\text{max,norm}}/M, P_{\text{avg,norm}}) \leq P_{\text{avg,norm}}$.

After passing through the optical wireless channel, the signal is distorted by AWGN, $\mathbf{w}_l$, at the receiver to obtain the received signal, $\mathbf{y}_l$. After synchronization, the signal is passed through a filter with an impulse response of $v(\tau - t)$, matched to the impulse response of the pulse shaping filter at the transmitter in order to maximize the received SNR. At the analog-to-digital converter (ADC), the signal is sampled at a frequency of M/T_s [21, 45]. After equalization of the channel effect by means of linear FFE or nonlinear DFE with ZF or MMSE criteria, the distorted replica of the transmitted symbol, $\tilde{s}_l$, is obtained. A hard-decision or soft-decision decoder can be employed to obtain the received bits.

The BER system performance in the optical wireless channel with AWGN and equalization has been discussed in [21]. The received M-PPM symbol can be treated as an OOK sequence, and the information bits can be decoded by means of a hard-decision decoder. For this approach, an analytical union bound of the BER as a function of the electrical SNR per bit is presented and verified through simulation. Alternatively, the BER performance can be enhanced by means of soft-decision decoding based on the position of the chip with the maximum level within the received M-PPM symbol. The analytical BER performance of this decoder is presented in [45] as an integral that cannot be derived in closed form. In addition, it requires a considerable computational effort to evaluate it for higher-order M-PPM. A tighter union bound for the symbol-error rate (SER) in soft-decision decoding can be obtained as a summation of the probabilities of $M-1$ chips of the total M chips in the M-PPM symbol being greater than an intended chip, c_1. Since chips in the chip vector, $\mathbf{c}$, are equally probable, a union bound for the SER can be expressed as follows:

$$\text{SER} \leq (M-1)P(c_2|c_1) = (M-1)\int_{-\infty}^{\infty} P(c_2|c_1 = s)P(c_1 = s)\,ds$$

$$= (M-1)\int_{-\infty}^{\infty} Q\left(\frac{s\sqrt{2}}{\sqrt{N_0}}\right)\frac{1}{\sqrt{\pi N_0}}$$

$$\times \exp\left(-\frac{\left(s - \sqrt{G_{\text{EQ}}\log_2(M)E_{b(\text{elec})}}\right)^2}{N_0}\right)ds\,,$$
(4.4)

where $Q(\cdot)$ is the complementary cumulative distribution function (CCDF) of a standard normal distribution with zero-mean and unity variance. The BER can be obtained as follows:

$$\text{BER} = \frac{M\log_2(M)}{2(M-1)}\text{SER}\,.$$
(4.5)

4.3.2   Pulse amplitude modulation: M-PAM

The block diagram of M-PAM is presented in Fig. 4.4. Here, $\log_2(M)$ equiprobable input bits form a time-domain symbol with a bandwidth of $B = 1/T_s$ and a duration of T_s for a symbol rate of $R_s = B$. The symbols are assigned to current levels of $p\sqrt{3P_{s(\text{elec})}}/\sqrt{(M-1)(M+1)}$, $p = \pm 1, \pm 3, \ldots, \pm(M-1)$, and these are grouped in the train of L symbols, $\mathbf{s}_l$. Here, $E_{b(\text{elec})} = P_{b(\text{elec})}/B = P_{s(\text{elec})}/(\log_2(M)B)$. The resulting spectral efficiency of M-PAM is $\log_2(M)$ bits/s/Hz [21, 45]. The train of symbols is predistorted, scaled, and passed through the DAC. A pulse shaping filter is applied. Since s_l is bipolar, it requires a DC bias, β_{DC}, to fit within the dynamic range of the front-end. The transmitted signal vector, $F(\mathbf{x}_l)$, has a length of $Z_\mathbf{x} = L$, and it can be expressed as follows:

$$F(\mathbf{x}_l) = \alpha \mathbf{s}_l + \beta_{DC} , \tag{4.6}$$

where

$$\alpha = \sqrt{\frac{M+1}{3(M-1)P_{s(elec)}}} \min\left(P_{max,norm} - \beta_{DC}, \beta_{DC} - P_{min,norm}\right) . \tag{4.7}$$

In order to obtain the O/E conversion in M-PAM from (4.3), the second moment of $F(\mathbf{x}_l)$ can be expressed as follows:

$$\mathrm{E}\left[F(\mathbf{x}_l)^2\right] = \frac{M+1}{3(M-1)} \min\left(P_{max,norm} - \beta_{DC}, \beta_{DC} - P_{min,norm}\right)^2 + \beta_{DC}^2 . \tag{4.8}$$

Because of the fact that the M-PAM time-domain signal has a PDF with finite support, it can be fitted within $P_{min,norm}$ and $P_{max,norm}$ without clipping. Thus, the following holds for its average optical power: $\mathrm{E}[F(\mathbf{x}_l)] = P_{s(opt)} = \beta_{DC} \leq P_{avg,norm}$.

The signal is transmitted over the optical wireless channel, and it is distorted by AWGN, $\mathbf{w}_l$, at the receiver. The received signal, $\mathbf{y}_l$, is synchronized, and it is passed through a matched filter. At the ADC, the signal is sampled at a frequency of $1/T_s$ [21, 45]. Next, the signal is equalized by means of linear FFE or non-linear DFE with ZF or MMSE criteria, and the distorted replica of the transmitted symbol, $\tilde{s}_l$, is obtained. A hard-decision decoder is employed to obtain the received bits. As a result, the effective electrical SNR per bit in M-PAM, $\Gamma_{b(elec)}$, can be expressed as a function of the undistorted electrical SNR per bit at the transmitter, $\gamma_{b(elec)}$, as follows:

$$\Gamma_{b(elec)} = \gamma_{b(elec)} G_{EQ} G_{DC} . \tag{4.9}$$

Here, the equalizer gain factor, G_{EQ}, is given in Chapter 2. The gain factor G_{DC} denotes the attenuation of the useful electrical signal power of $\mathbf{x}$ due to the DC component. In general, the addition of the DC bias influences the useful electrical power of the biased time-domain signal. The total electrical power is a summation of the useful electrical alternating current (AC) power and the electrical DC power. Therefore, for a fixed total electrical power, the addition of the DC bias reduces the useful electrical AC power of the signal. The gain factor G_{DC} is given for M-PAM as follows:

$$G_{DC} = \frac{\mathrm{E}\left[(F(\mathbf{x}_l) - F(\beta_{DC}))^2\right]}{\mathrm{E}\left[F(\mathbf{x}_l)^2\right]} , \tag{4.10}$$

where

$$\mathrm{E}\left[(F(\mathbf{x}_l) - F(\beta_{DC}))^2\right] = \frac{M+1}{3(M-1)} \min\left(P_{max,norm} - \beta_{DC}, \beta_{DC} - P_{min,norm}\right)^2 . \tag{4.11}$$

The exact closed-form expression for the BER performance of M-PAM in AWGN has been presented in [160] as a summation of M terms. A tight approximation for BER below 10^{-2} can be obtained by considering the error contributed only by the closest symbols in the constellation as follows [45]:

$$\text{BER} = \frac{N_s}{G_{GC} \log_2(M)} Q\left(d_s \sqrt{\frac{\log_2(M)}{2}} \Gamma_{b(elec)}\right). \tag{4.12}$$

In M-PAM, an intended symbol has an average number of $N_s = 2(M-1)/M$ neighboring symbols. The gain introduced by Gray coding of the bits on the symbols is denoted by $G_{GC} = 1$. The distance between an intended symbol and the closest interfering symbol is given by $d_s = \sqrt{12/(M^2 - 1)}$.

4.3.3   BER performance with pre-distortion in AWGN

The multiplication of the bandwidth and the spectral efficiency of the OWC systems yields the system throughput. However, for a fixed SNR, the BER performance is independent of the bandwidth, and consequently of the throughput. In this study, the BER performance of the OWC systems is compared for equal average electrical symbol power, $P_{s(elec)}$, and equal bandwidth, B. In addition, the BER is assessed through the effective electrical SNR per bit at the receiver as a function of the undistorted electrical SNR per bit at the transmitter, i.e. the average electrical bit energy normalized to the power spectral density of the AWGN, $\gamma_{b(elec)} = E_{b(elec)}/N_0$.

Link parameters such as the optical center frequency, the mutual orientation and position of the transmitter and receiver in an indoor setup and their field of view (FOV), the responsivity and the photosensitive area of the detector, and the average radiated optical power of the transmitter determine the optical path gain coefficient, $g_{h(opt)}$ [16, 17], and, therefore, the electrical path gain coefficient, $g_{h(elec)}$, from (2.17). Since, however, $g_{h(elec)}$ is merely a factor in the equalization process, a change in $g_{h(elec)}$ results in an equal SNR penalty for all the considered systems. Therefore, $g_{h(elec)} = 1$ is assumed for simplicity. In general, the maximum information rate of the OWC system is achieved when the coherence bandwidth of the channel is significantly larger than the signal bandwidth, i.e. when the channel is considered as flat fading with an impulse response of $h(t) = \delta(t)$, where $\delta(t)$ is the Dirac delta function. Equivalently, the maximum information rate is achievable for a channel frequency response of $|H(f)|^2 = 1$. The BER performance of the OWC systems in AWGN is shown here for the flat fading channel without dispersion. An elaborate study of the system performance in a dispersive channel is presented in Chapter 5. In addition, a perfect synchronization between the transmitter and the receiver is assumed in the processing of the information-carrying signal. Moreover, perfect matched filtering and perfect channel knowledge at the receiver and the transmitter are assumed for the purposes of equalization and bit and power loading.

The analytical expressions for the BER performance of M-PPM and M-PAM in AWGN are compared with a Monte Carlo BER simulation. The following modulation orders are chosen: $M = \{2, 4, 16, 64, 256, 1024\}$. Signal pre-distortion is applied at the transmitter, in order to linearize its dynamic range of signal transfer. The practical 10-dB linear dynamic range of a Vishay TSHG8200 LED at room temperature ranged between 5 mW and 50 mW of radiated optical power [91]. Therefore, the linear dynamic range of the optical front-end is assumed to be between $P_{\text{min,norm}} = 0.1$ and $P_{\text{max,norm}} = 1$. The transmitted signal spans the entire dynamic range of optical power and no constraint is imposed on the radiated average optical power, in order to obtain the best BER system performance for the given dynamic range. The average optical power level is set in M-PPM to $\mathrm{E}\,[F(\mathbf{x}_l)] = P_{\text{max,norm}}/M$, and in M-PAM to $\mathrm{E}\,[F(\mathbf{x}_l)] = (P_{\text{min,norm}} + P_{\text{max,norm}})/2$.

The BER performance of M-PPM as a function of the electrical SNR per bit at the transmitter, i.e. $\gamma_{\text{b(elec)}} = E_{\text{b(elec)}}/N_0$, is presented in Fig. 4.5. As expected, the union bound is tight at low modulation orders with a 0.3-dB gap at 10^{-3} BER for 1024-PPM. It is shown that M-PPM is a suitable modulation technique for the low-rate low-SNR regime in the flat fading channel, even without forward error correction (FEC) coding. The BER performance of M-PAM is presented in Fig. 4.6. The theoretical and simulation results exhibit a very close match. Since the standard PAM symbol is bipolar, a DC bias is required to fit the signal within the dynamic range of the transmitter. Due to the DC-bias penalty, G_{DC}, optical M-PAM has an increased electrical SNR requirement. If the electrical power budget is sufficient to accommodate a higher modulation order, M-PAM is a suitable candidate for high-rate data transmission in the high-SNR regime of the flat fading channel.

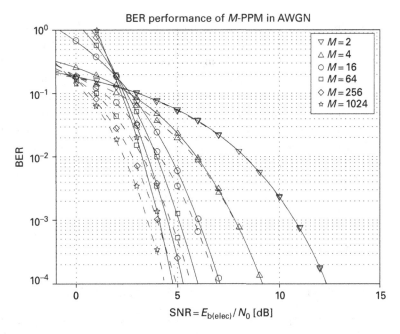

Figure 4.5 BER performance of M-PPM in AWGN for a 10-dB dynamic range, simulation (dashed lines) vs. theory (solid lines)

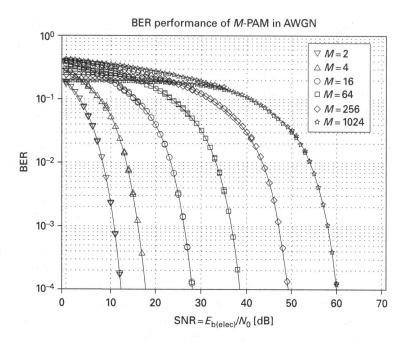

Figure 4.6 BER performance of *M*-PAM in AWGN for a 10-dB dynamic range, simulation (dashed lines) vs. theory (solid lines)

4.4   Multi-carrier modulation

In multi-carrier modulation, the information is encoded in complex-valued subcarriers with *M*-QAM as the underlying modulation scheme. The IFFT and FFT are utilized as multiplexing and demultiplexing techniques at the transmitter and receiver, respectively. Through Hermitian symmetry on the OFDM frame, a real-valued time-domain signal is obtained at the expense of 50% reduction of the spectral efficiency. O-OFDM transmission is generally realized by a DC-biased Gaussian signal, i.e. DCO-OFDM, or a half-Gaussian signal with a further 50% reduction of the data rate, i.e. ACO-OFDM. The signal biasing setup is defined by signal scaling and DC-biasing in order to condition the DCO-OFDM and ACO-OFDM time-domain signals within the positive dynamic range of the transmitter. The received electrical SNR at an information-carrying subcarrier is derived, including the non-linear distortion parameters, such as the attenuation factor and the variance of the non-linear distortion noise, for any given front-end biasing setup. The BER performance of the O-OFDM schemes is presented and verified by means of a Monte Carlo simulation.

4.4.1   Optical OFDM with *M*-QAM: DCO-OFDM and ACO-OFDM

The block diagram for multi-carrier O-OFDM transmission is presented in Fig. 4.7. O-OFDM transmission can be realized through the bipolar DCO-OFDM with a DC bias [147] or through the unipolar ACO-OFDM [51]. In general, *N* subcarriers form

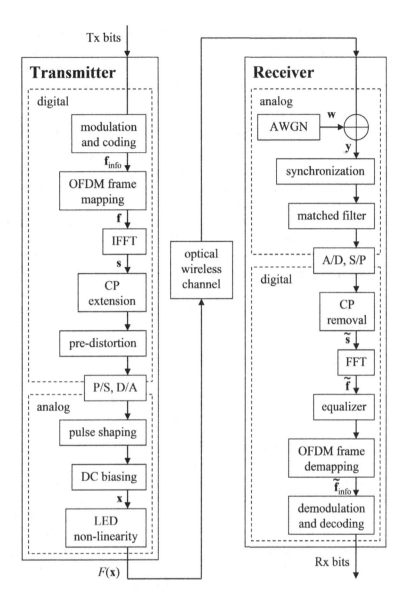

Figure 4.7 Block diagram of multi-carrier transmission in OWC using OFDM

the lth OFDM frame, $\mathbf{f}_{l,m}$, corresponding to the lth OFDM symbol, where m, $m = 0$, $1, \ldots, N-1$, is the subcarrier index. Each subcarrier occupies a bandwidth of $1/NT_s$ in a total OFDM frame bandwidth of $B = 1/T_s$. The two O-OFDM systems utilize a different portion of the available bandwidth, and the bandwidth utilization factor is denoted by G_B, where $G_B = (N-2)/N$ in DCO-OFDM and $G_B = 0.5$ in ACO-OFDM. In order to ensure a real-valued time-domain signal, both schemes have Hermitian symmetry imposed on the OFDM frame, and the subcarriers with indices $m = \{0, N/2\}$ are set to zero. In DCO-OFDM, $N/2 - 1$ subcarriers in the first half of the frame, $m = 1, \ldots, N/2 - 1$, carry the information. In ACO-OFDM, only the odd subcarriers are enabled, while every even subcarrier is set to zero. Here, the $N/4$ subcarriers with indices $m = 1, 3, 5, \ldots, N/2 - 1$ carry the information. The information-carrying subcarriers are denoted as $\mathbf{f}_{\text{info}}^{l,m}$. In order to maintain the signal variance of σ^2, the electrical power of the enabled subcarriers is scaled by $\alpha = \sigma/\sqrt{G_B}$ to $P_{\text{s(elec)}}/G_B$, where $P_{\text{s(elec)}} = \sigma^2$. Both schemes can utilize bit and power loading of the frequency domain subcarriers in order to optimally adapt the signal to the channel conditions. For a desired bit rate per unit bandwidth, the Levin–Campello algorithm [49, 50] can be applied in order to maximize the received power margin, or, equivalently, in order to minimize the required electrical SNR. The optimum solution achieved by the algorithm yields the number of bits, $\mathbf{b}_m$, which modulate a complex-valued information-carrying frequency domain subcarrier from $\mathbf{f}_{l,m}$ using M-QAM. In addition, the algorithm provides subcarrier power scaling factors, $\mathbf{p}_m$, which ensure an equal maximized received power margin for every active subcarrier. Therefore, $\mathrm{E}\,[\mathbf{b}_m]\,G_B N$ bits on the $G_B N$ enabled subcarriers are transmitted per OFDM time-domain symbol, $\mathbf{s}_l$. Without loss of generality, only integer average bit load, i.e. $\mathrm{E}\,[\mathbf{b}_m] = \log_2(M)$, is considered in this study. Thus, the average bit energy can be expressed as follows: $E_{\text{b(elec)}} = P_{\text{b(elec)}}/(G_T G_B B) = \sigma^2/(\log_2(M) G_T G_B B)$, where $P_{\text{b(elec)}}$ is the average electrical power of $G_B N$ bits. The lth OFDM symbol in the train of L symbols, $\mathbf{s}_l$, is obtained by the IFFT of the lth OFDM frame in the train of L frames, $\mathbf{f}_l$. Here, the unitary IFFT is utilized as a multiplexing technique at the transmitter, as follows:

$$s_{l,k} = \frac{1}{\sqrt{N}} \sum_{m=0}^{N-1} \mathbf{f}_{l,m} \exp\left(\frac{i2\pi km}{N}\right), \tag{4.13}$$

where $i = \sqrt{-1}$ is the imaginary unit, and k, $k = 0, 1, \ldots, N - 1$, denotes the time-domain sample index within the lth OFDM symbol. In addition, N_{CP} samples from the end of each OFDM symbol are appended at the beginning of the symbol, creating the CP extension in order to mitigate the ISI and the inter-carrier interference (ICI) [99, 107]. A large number of subcarriers and a CP transform the dispersive optical wireless channel into a flat fading channel over the subcarrier bandwidth, reducing the computational complexity of the equalization process at the receiver to a single-tap equalizer [99]. As a result, the time-domain OFDM symbol with CP occupies a bandwidth of $B = 1/T_s$, and it has a duration of $(N + N_{\text{CP}})T_s$. The symbol rate amounts to $R_s = G_T G_B B$ [99, 107], where $G_T = N/(N + N_{\text{CP}})$ is the utilization factor for the information-carrying time. Because of Hermitian symmetry, the resulting spectral efficiency of M-QAM O-OFDM amounts to $\log_2(M) G_T G_B/2$ bits/s/Hz. The train of OFDM symbols with CPs follows a close to Gaussian distribution with zero-mean and variance of σ^2 for FFT sizes greater than 64 [110]. The non-linear transfer characteristic of the LED transmitter can be compensated by pre-distortion with the inverse of the non-linear transfer function [135]. The pre-distorted OFDM symbol is subjected to a parallel-to-serial (P/S) conversion, and it is passed through the DAC to obtain the continuous-time signal. A pulse shaping filter is applied. Since only the odd subcarriers are enabled in ACO-OFDM, the negative portion of the OFDM symbol can be clipped without loss of information at the received odd subcarriers. The signal is DC-biased by β_{DC} in the analog circuitry, and it is transmitted by the LED. The transmitted signal vector, $F(\mathbf{x}_l)$, with a length of $Z_\mathbf{x} = L(N + N_{CP})$ can be expressed as follows:

$$F(\mathbf{x}_l) = \Phi(\mathbf{s}_l + \beta_{DC}) , \qquad (4.14)$$

where $\Phi(\cdot)$ represents the linearized transfer function of the LED transmitter which is reduced to the double-sided signal clipping operator discussed in Chapter 3. Effectively, the OFDM symbol, $\mathbf{s}_l$, is clipped at normalized bottom and top clipping levels of λ_{bottom} and λ_{top} relative to a standard normal distribution [131]. In DCO-OFDM, $\lambda_{bottom} = (P_{min,norm} - \beta_{DC})/\sigma$, while in ACO-OFDM, $\lambda_{bottom} = \max((P_{min,norm} - \beta_{DC})/\sigma, 0)$. In both systems, $\lambda_{top} = (P_{max,norm} - \beta_{DC})/\sigma$. The clipping levels in DCO-OFDM can be negative and/or positive, whereas in ACO-OFDM, these are strictly non-negative. For reasons of plausibility, $\lambda_{top} > \lambda_{bottom}$. The Bussgang theorem [137] and the CLT [136] can be used to translate the non-linear distortion into an attenuation factor for the information-carrying subcarriers, K, plus uncorrelated zero-mean complex-valued Gaussian noise with variance of σ_{clip}^2. The details of the analysis of these non-linear distortion parameters have been presented in Section 3.4 of Chapter 3.

In addition to the distortion of the information-carrying subcarriers, the time-domain signal clipping modifies the average optical power of the transmitted signal. The modified mean can be derived based on the statistics of a truncated Gaussian distribution [161]. Thus, the average optical power of the transmitted signal after double-sided signal clipping, $E[\Phi(\mathbf{x}_l)]$ can be expressed as follows:

$$E[\Phi(\mathbf{x}_l)] = \sigma\left(\lambda_{top}Q(\lambda_{top}) - \lambda_{bottom}Q(\lambda_{bottom}) + \phi(\lambda_{bottom}) - \phi(\lambda_{top})\right) + P_{bottom} .$$
$$(4.15)$$

In DCO-OFDM, $P_{bottom} = P_{min,norm}$. Because of the default zero-level clipping in ACO-OFDM, $P_{bottom} = \max(P_{min,norm}, \beta_{DC})$. Here, $\phi(\cdot)$ denotes the PDF of a standard normal distribution with zero-mean and unity variance. The eye safety regulations [44] and/or the design requirements such as signal dimming impose the average optical power constraint, $P_{avg,norm}$, i.e. $E[\Phi(\mathbf{x}_l)] \leq P_{avg,norm}$. Thus, for a given set of front-end optical power constraints, it is possible to obtain the target signal variance, σ^2, and the required DC bias, β_{DC}, from (4.15). The optimum choice of these design parameters is elaborated in Chapter 5.

Since the signal is clipped, the resulting average optical power of the signal, $E[\Phi(\mathbf{x}_l)]$, differs from the undistorted optical power of the OFDM symbol, $P_{s(opt)}$. In general, $P_{s(opt)}$ is defined for the scenario with the least signal clipping which is given in DCO-OFDM as $\lambda_{bottom} = -\infty$ and $\lambda_{top} = +\infty$, and in ACO-OFDM as $\lambda_{bottom} = 0$ and $\lambda_{top} = +\infty$. In DCO-OFDM, $P_{s(opt)} = \beta_{DC}$, whereas in ACO-OFDM, $P_{s(opt)} = (\beta_{DC} + \sigma/\sqrt{2\pi})$. The O/E conversion is obtained in DCO-OFDM and ACO-OFDM, respectively, as follows:

$$P_{s(elec)} = \frac{\sigma^2 + \beta_{DC}^2}{\beta_{DC}^2} P_{s(opt)}^2 , \qquad (4.16)$$

$$P_{s(elec)} = \frac{2\pi\beta_{DC}^2 + 2\sigma\beta_{DC}\sqrt{2\pi} + \pi\sigma^2}{2\pi\beta_{DC}^2 + 2\sigma\beta_{DC}\sqrt{2\pi} + \sigma^2} P_{s(opt)}^2 . \qquad (4.17)$$

The signal $\mathbf{x}_{l,k}$ is transmitted over the optical wireless channel, and it is distorted by AWGN, $\mathbf{w}_{l,k}$, at the receiver. The received signal, $\mathbf{y}_{l,k}$, is synchronized, and it is passed through a matched filter. At the ADC, the signal is sampled at a frequency of $1/T_s$ [99, 107]. After serial-to-parallel (S/P) conversion, the CP extension of each OFDM symbol is removed to obtain the distorted replica of the OFDM symbol, $\tilde{\mathbf{s}}_{l,k}$. Next, the signal is passed through an FFT block back to the frequency domain. Here, the demultiplexing is carried out by the use of the unitary FFT, as follows:

$$\tilde{\mathbf{f}}_{l,m} = \frac{1}{\sqrt{N}} \sum_{k=0}^{N-1} \tilde{\mathbf{s}}_{l,k} \exp\left(\frac{-i2\pi km}{N}\right), \tag{4.18}$$

where $\tilde{\mathbf{f}}_{l,m}$ denotes the distorted replica of the OFDM frame, respectively. Here, the CLT can be applied, and the time-domain additive uncorrelated non-Gaussian non-linear noise is transformed into additive uncorrelated zero-mean complex-valued Gaussian noise at the information-carrying subcarriers, $\mathbf{W}_{\text{clip}}^{l,m}$, preserving its variance of σ_{clip}^2. Because of the fact that the channel frequency response can be considered as flat fading over the subcarrier bandwidth, the signal can be equalized by means of a single-tap linear FFE with ZF or MMSE criteria. As a result, the received distorted replica of the information-carrying subcarriers, $\tilde{\mathbf{f}}_{\text{info}}^{l,m}$, can be expressed in DCO-OFDM and ACO-OFDM as follows:

$$\tilde{\mathbf{f}}_{\text{info}}^{l,m} = \left(\sqrt{\mathbf{p}_{l,m}} K \mathbf{f}_{\text{info}}^{l,m}/\sqrt{G_B} + \mathbf{W}_{\text{clip}}^{l,m}\right) \sqrt{G_{\text{EQ}} G_{\text{DC}}} + \sqrt{G_B} \mathbf{W}_{l,m}. \tag{4.19}$$

Here, G_{EQ} simplifies for the single-tap linear FFE with ZF or MMSE criteria, respectively, to $|H(f_{\text{info}})|^2$ and $|H(f_{\text{info}})|^2 + \gamma_{\text{b(elec)}}^{-1}$, where $H(f_{\text{info}})$ represents the channel frequency response on the information-carrying subcarrier. The zero-mean complex-valued AWGN with variance of $\sigma_{\text{AWGN}}^2 = BN_0$ is given by $\mathbf{W}_{l,m}$. The gain factor G_{DC} denotes the attenuation of the useful electrical signal power due to the DC component, i.e. the penalty on the electrical SNR due to the DC component, and it can be expressed in DCO-OFDM and ACO-OFDM, respectively, as follows:

$$G_{\text{DC}} = \frac{\sigma^2}{\sigma^2 + \beta_{\text{DC}}^2}, \tag{4.20}$$

$$G_{\text{DC}} = \frac{\sqrt{2\pi}\sigma^2}{\sqrt{2\pi}\sigma^2 + 4\sigma\beta_{\text{DC}} + 2\sqrt{2\pi}\beta_{\text{DC}}^2}. \tag{4.21}$$

The generalized model of the received information-carrying subcarriers in O-OFDM is illustrated in Fig. 4.8. A hard-decision decoder is employed on the known OFDM frame structure to obtain the received bits. Thus, the electrical SNR on a received information-carrying subcarrier in DCO-OFDM and ACO-OFDM is given as follows:

$$\text{SNR}_m = \frac{\mathbf{p}_m K^2 P_{\text{s(elec)}}/G_B}{\sigma_{\text{clip}}^2 + \frac{G_B \sigma_{\text{AWGN}}^2}{G_{\text{EQ}} G_{\text{DC}}}}. \tag{4.22}$$

Considering that $P_{\text{b(elec)}} = E_{\text{b(elec)}} G_T G_B B$, the effective electrical SNR per bit on an enabled subcarrier in O-OFDM, $\Gamma_{\text{b(elec)}}^m$, can be expressed as a function of the

Table 4.1 Parameters in (4.12) for square and cross M-QAM [45, 160, 162, 163]

Modulation order	N_s	G_{GC}	d_s
BPSK	1	1	2
$M = 2^{2n}$ $n = 1, 2, \ldots$	$4 - \frac{4}{\sqrt{M}}$	1	$\sqrt{\frac{6}{M-1}}$
$M = 8$	3	$\frac{4}{5}$	$\sqrt{\frac{4}{3+\sqrt{3}}}$
$M = 32$	$\frac{13}{4}$	$\frac{6}{7}$	$\frac{1}{\sqrt{5}}$
$M = 2^{2n+1}$ $n = 3, 4, \ldots$	$4 - \frac{6}{\sqrt{2M}}$	$\frac{6M}{6M+3\sqrt{2M}+2}$	$\sqrt{\frac{192}{31M-32}}$

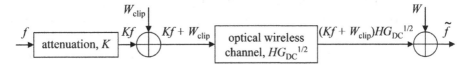

Figure 4.8 Generalized block diagram of the received information-carrying subcarriers in O-OFDM. It includes the attenuation factor and the additive Gaussian noise from transmitter non-linearity, the channel tap, the DC-bias penalty, and the AWGN

undistorted electrical SNR per bit at the transmitter, $\gamma_{b(elec)}$, as follows:

$$\Gamma^m_{b(elec)} = \frac{\mathbf{p}_m K^2 P_{b(elec)}/G_B}{\sigma^2_{clip} + \frac{G_B \sigma^2_{AWGN}}{G_{EQ}G_{DC}}} = \frac{\mathbf{p}_m K^2}{\mathbf{b}_m G_B \sigma^2_{clip} + \frac{G_B \gamma^{-1}_{b(elec)}}{G_T G_{EQ}G_{DC}}} . \tag{4.23}$$

The exact closed-form expression for the BER performance of square and cross M-QAM constellations in AWGN has been presented as a summation of M terms in [160] and [162], respectively. However, the same tight approximation from (4.12) can be applied, and the respective parameters are given in Table 4.1 [45, 160, 162, 163], including binary phase shift keying (BPSK). Thus, the BER on the intended subcarrier, BER_m, can be obtained by inserting (4.23) into (4.12), considering the parameters from Table 4.1 for the number of loaded bits. As a result, the link BER can be obtained as the average of the BER of all enabled subcarriers: $BER = E[BER_m]$.

4.4.2 BER performance with generalized non-linear distortion in AWGN

The accuracy of the non-linear distortion modeling and the derived expression for the electrical SNR at the received subcarriers in (4.23) is verified by means of a Monte Carlo BER simulation. For this purpose, an FFT size of 2048 is used to ensure Gaussianity of the O-OFDM signals. A set of QAM orders, $M = \{16, 64, 256\}$, is chosen to test the model with lower and higher modulation orders. The BER performance of DCO-OFDM and ACO-OFDM is evaluated as a function of the undistorted electrical SNR per bit, $\gamma_{b(elec)}$. The analytical expression for the BER from (4.12) and Table 4.1 is used. The validity of the model is presented for the general non-linear piecewise polynomial transfer function, $F(x) = \Xi(x)$, considered in $\Psi(s) = F(x) - F(\beta_{DC})$. An example for $\Xi(x)$, illustrated in Fig. 3.3, with increased precision of the polynomial coefficients for the purposes of the accurate model verification is chosen as follows:

$$\Xi(x) = \begin{cases} 0 & \text{if } x \le 0.1, \\ -1.6461x^3 + 2.7160x^2 \\ -0.4938x + 0.0239 & \text{if } 0.1 < x \le 1, \\ 0.6 & \text{if } x > 1. \end{cases} \tag{4.24}$$

To condition the signal within this non-linear transfer function, the front-end biasing setup is defined by the DC bias, β_{DC}, and the signal standard deviation, σ. In DCO-OFDM, $\beta_{DC} = 0.5$ and $\sigma = 0.07$, while in ACO-OFDM, $\beta_{DC} = 0.4$ and $\sigma = 0.24$. This setup results in an equal radiated average optical power of 0.25 for both O-OFDM schemes. It enables ACO-OFDM to avoid the bottom knee of the non-linear transfer function and, therefore, to reduce distortion of the half-Gaussian signal for the objective of better BER performance. In DCO-OFDM, the signal is placed below the middle of the dynamic range, as suggested in [43], in order to improve electrical power efficiency. In addition, this setup exhibits a moderate attenuation factor, K, and non-linear noise variance, σ_{clip}^2, in order to validate the accuracy of the non-linear distortion model against higher-order modulation. The BER performance of the DCO-OFDM and ACO-OFDM systems is presented in Fig. 4.9. It is ascertained that the theoretical and

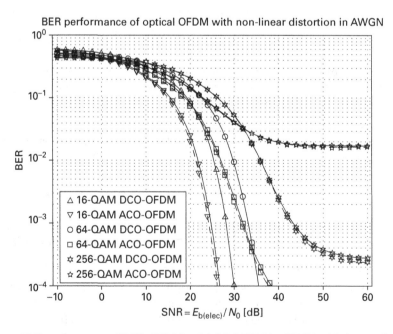

Figure 4.9 BER performance of DCO-OFDM and ACO-OFDM in AWGN with the non-linear distortion function $\Xi(x)$, simulation (dashed lines) vs. theory (solid lines) [133]. ©2013 IEEE

simulation results confirm a close match. Since only the odd subcarriers are modulated in ACO-OFDM, the electrical SNR requirement of M-QAM DCO-OFDM has to be compared with the one of M^2-QAM ACO-OFDM for an equal information rate. It is shown that ACO-OFDM experiences greater BER degradation, even though the bottom knee of the non-linear transfer function is avoided. DCO-OFDM consistently demonstrates a lower electrical SNR requirement as compared to ACO-OFDM for modulation orders with equal information rate, while higher-order modulation proves to be more vulnerable to non-linear signal distortion.

4.4.3 BER performance with pre-distortion in AWGN

The accuracy of the analysis of the double-sided signal clipping distortion in O-OFDM is verified by means of a Monte Carlo BER simulation. For this purpose, an FFT size of 2048 is chosen. The BER performance of DCO-OFDM and ACO-OFDM is evaluated as a function of the undistorted electrical SNR per bit, $\gamma_{b(\text{elec})}$, in (4.23). The analytical expression for the BER from (4.12) and Table 4.1 is used. In general, because of the structure of the OFDM frame, ACO-OFDM achieves half the spectral efficiency of DCO-OFDM for equal modulation orders. Therefore, in addition to 4-QAM ACO-OFDM with spectral efficiency of 0.5 bits/s/Hz, 16-QAM ACO-OFDM is compared with 4-QAM DCO-OFDM to evaluate the BER performance at a similar spectral efficiency of 1 bit/s/Hz. Furthermore, the systems are constrained for equal average optical power of the transmitted signal. The linearized transfer characteristic with double-sided clipping, $F(x) = \Phi(x)$, is considered in $\Psi(s) = F(x) - F(\beta_{\text{DC}})$. Here, a practical 10-dB linear dynamic range of a Vishay TSHG8200 LED at room temperature, i.e. $P_{\text{min,norm}} = 0.1$ and $P_{\text{max,norm}} = 1$, is considered at the transmitter [91]. No clipping at the receiver is assumed. As benchmarks for a comparison with existing results, idealized cases for signal scaling and biasing are included from [138]. For any given fixed average optical power at the transmitter, $\text{E}[\Phi(x)] = P_{\text{avg,norm}}$, the ACO-OFDM signal is clipped in the idealized case at $\lambda_{\text{bottom}} = 0$ and $\lambda_{\text{top}} = +\infty$. Here, $\beta_{\text{DC}} = 0$ and $\sigma = P_{\text{avg,norm}}\sqrt{2\pi}$. In DCO-OFDM, a widely used idealized case is defined for $\lambda_{\text{bottom}} = -2$ and $\lambda_{\text{top}} = +\infty$ [138]. Here, $\beta_{\text{DC}} = P_{\text{avg,norm}}$ and $\sigma = P_{\text{avg,norm}}/2$. In addition to the idealized case, two clipping scenarios which satisfy the average optical power constraint are considered for the verification of the analytical framework. Two average optical power constraints are chosen, i.e. $\text{E}[\Phi(x)] = P_{\text{avg,norm}} = 0.2$ and $\text{E}[\Phi(x)] = P_{\text{avg,norm}} = 0.3$. Here, the front-end biasing is defined through the parameters β_{DC} and σ. They are obtained from (4.15) as a pair which yields the chosen $P_{\text{avg,norm}}$ for the given $P_{\text{min,norm}}$ and $P_{\text{max,norm}}$. In the first case, $P_{\text{avg,norm}} = 0.2$ is realized in ACO-OFDM for $\beta_{\text{DC}} = 0.08$ and $\sigma = 0.29$, whereas $\beta_{\text{DC}} = 0.2$ and $\sigma = 0.1$ are required in DCO-OFDM. In the second case, $P_{\text{avg,norm}} = 0.3$ is obtained in ACO-OFDM for $\beta_{\text{DC}} = 0.06$ and $\sigma = 0.61$, while $\beta_{\text{DC}} = 0.3$ and $\sigma = 0.15$ are considered in DCO-OFDM. The two biasing setups yield the following normalized clipping levels. In ACO-OFDM, $\lambda_{\text{bottom}} = 0.06$ and $\lambda_{\text{top}} = 3.15$ in the first case, whereas $\lambda_{\text{bottom}} = 0.07$ and $\lambda_{\text{top}} = 1.54$ in the second case. In DCO-OFDM, $\lambda_{\text{bottom}} = -0.98$ and $\lambda_{\text{top}} = 8.2$ in the first case, whereas $\lambda_{\text{bottom}} = -1.32$ and $\lambda_{\text{top}} = 4.76$ in the second case.

ACO-OFDM is expected to perform better in the first biasing setup because of the less severe upside clipping of the half-Gaussian signal distribution. On the contrary, DCO-OFDM is expected to perform better in the second biasing setup because the normalized clipping levels are closer to a symmetric clipping of the Gaussian signal distribution.

The BER performance of the studied ACO-OFDM and DCO-OFDM systems is presented in Figs. 4.10 and 4.11. The theoretical and simulation results confirm a very close match. It is shown that the existing simulation results from [138] for an idealized biasing scenario underestimate the BER performance of ACO-OFDM and DCO-OFDM in a biasing setup with a clipping distortion. However, the idealized biasing scenarios are not achievable in practice due to the limitations of off-the-shelf components in terms of electrical and optical power constraints. In general, for identical QAM modulation orders ACO-OFDM demonstrates a better BER performance as compared to DCO-OFDM at the expense of 50% reduction in spectral efficiency. In Fig. 4.10, even though 4-QAM ACO-OFDM shows the better BER performance for a corresponding clipping scenario, it is outperformed by 4-QAM DCO-OFDM in terms of spectral efficiency over the entire SNR region [164]. However, the results show that the biasing setup plays a significant role when comparing 16-QAM ACO-OFDM and 4-QAM DCO-OFDM. In general, a setup which accommodates lower average optical power with

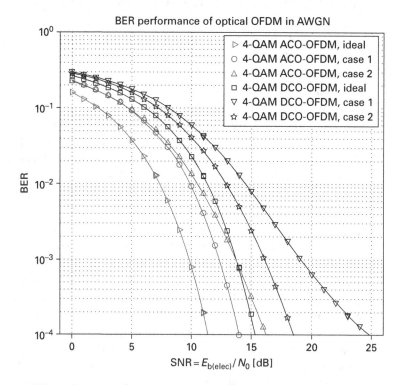

Figure 4.10 BER performance of 4-QAM ACO-OFDM and 4-QAM DCO-OFDM in AWGN for a 10-dB linear dynamic range, simulation (dashed lines) vs. theory (solid lines) [131]. ©2012 IEEE

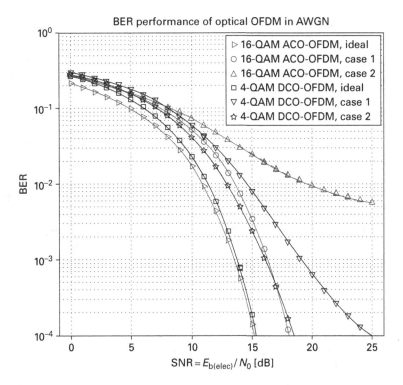

Figure 4.11 BER performance of 16-QAM ACO-OFDM and 4-QAM DCO-OFDM in AWGN for a 10-dB linear dynamic range, simulation (dashed lines) vs. theory (solid lines) [131]. ©2012 IEEE

respect to the given dynamic range is more suitable for ACO-OFDM. However, shifting the average optical power towards the middle of the dynamic range can potentially yield better BER performance with DCO-OFDM and thereby higher system throughput. This is because of the resulting front-end-induced signal clipping: in the first case DCO-OFDM is clipped more severely, whereas in the second case ACO-OFDM is the one that experiences greater signal degradation. This is confirmed by the BER results of 16-QAM ACO-OFDM and 4-QAM DCO-OFDM in Fig. 4.11. Whereas the first case of 16-QAM ACO-OFDM and the second case of 4-QAM DCO-OFDM perform similarly, the respective counterparts are more severely clipped, and their performance is degraded. However, a comparison between 4-QAM and 16-QAM ACO-OFDM suggests that higher order QAM modulation is more vulnerable to signal clipping. This is because, according to (4.23), the Gaussian clipping noise at the received data-carrying subcarriers is added independently of the QAM modulation order, M. Therefore, it is expected that M-QAM DCO-OFDM delivers a better BER performance, and therefore a higher throughput as compared to M^2-QAM ACO-OFDM in a biasing setup with a higher average optical power. In addition, since the factors in (4.23) are independent of N for practical FFT sizes, i.e. $N > 64$, the BER and throughput performance is independent of the total number of subcarriers.

4.5   Summary

In this chapter, single-carrier and multi-carrier modulation schemes for practical OWC links have been presented. The BER performance of the single-carrier M-PPM, and M-PAM, as well as the multi-carrier M-QAM DCO-OFDM and M-QAM ACO-OFDM, has been analyzed in detail. First, the structure of the different signals has been presented. Optical link impairments, such as the limited linear dynamic range of the transmitter, the dispersive optical wireless channel and the AWGN at the receiver, have been considered. The use of signal pre-distortion has been shown to linearize the dynamic range of the optical front-end between positive levels of minimum and maximum radiated optical power. In this way, non-linear distortion has been avoided in single-carrier pulse modulation schemes. In multi-carrier modulation schemes, non-linear distortion has been reduced to a double-sided signal clipping. In addition, average electrical power and average optical power constraints have been considered. The parameters of non-linear distortion in O-OFDM, such as the subcarrier attenuation factor and the variance of the non-linear noise component, have been included in the calculation of the received electrical SNR.

In the context of the positive limited dynamic range of the transmitter, the O/E conversion of the signals has been defined, taking into account the additional DC-bias requirement of some of the optical signals. In addition, the respective DC-bias penalties have been derived and included in the calculation of the electrical SNR. In order to counter the channel effect, equalization has been employed, and the equalizer penalty on the electrical SNR requirement has been considered. Finally, the electrical SNR has been translated into BER performance, and the analytical models have been verified by means of a Monte Carlo BER simulation.

The results show that the front-end biasing setup plays a significant role for the power efficiency of optical modulation schemes. In multi-carrier systems, reduction of the signal variance to avoid signal clipping leads to a large DC-bias penalty. LED non-linearity has also been shown to saturate the BER performance of DCO-OFDM and ACO-OFDM with higher order modulation, when excessive signal clipping is applied. Therefore, optimization of the biasing setup in the context of LED non-linearity is required in order to maximize the power efficiency. Linearizing the LED transfer function by pre-distortion helps to minimize the BER degradation. This technique can even completely eliminate the non-linear distortion in the case of single-carrier modulation due to confined signal distribution. While M-PPM can operate with a lower SNR requirement, it achieves this at the expense of a considerable reduction in spectral efficiency. While M-PAM can achieve high spectral efficiencies, it requires a very high SNR. Therefore, a comparison of the single-carrier and multi-carrier systems under similar electrical and optical power constraints is in order. The optimization of the front-end biasing setup and comparison of the power and spectral efficiencies of the optical modulation schemes is studied in detail in Chapter 5.

5 Spectral efficiency and information rate

5.1 Introduction

The capacity of a communication system in a given channel with noise is defined by Claude Shannon as the maximum number of bits per unit bandwidth that can be successfully transmitted (assuming an infinitesimally small error) over a communication link [165]. In a practical communication setup, the electrical signal-to-noise ratio (SNR) requirement for a target bit-error ratio (BER) performance in additive white Gaussian noise (AWGN), the corresponding spectral efficiency, and the information rate are figures of merit for a given modulation scheme. In this book, the spectral efficiency of the modulation scheme is referred to as the uncoded bit rate per unit bandwidth, while the information rate is based on the mutual information of transmitted and received data symbols. Therefore, the information rate represents the maximum achievable rate of the considered modulation scheme which in the given noisy channel can be practically obtained when symbol shaping and forward error correction (FEC) coding are applied [166]. In this context, the Shannon capacity of the communication system is the maximum of the mutual information between the transmitted and received signals in the given noisy channel, where the maximization is performed over all possible input signal distributions.

In a line-of-sight (LOS) communication scenario, where the optical wireless channel is dominated by the LOS signal component, the root-mean-squared (RMS) delay spread of the channel is very small, while the coherence bandwidth is very large, which enables the transmission of signals with equally broad information bandwidth, and very high data rates. Such a scenario can be generally described as a frequency non-selective slow fading channel or a flat fading channel. However, in a practical indoor setup, non-line-of-sight (NLOS) signal components can be reflected by the objects in a room and detected at the receiver. In this case, due to the multipath component, the RMS delay spread of the channel is increased, while the coherence bandwidth is reduced, and it is typically comparable or smaller than the bandwidth of the signal. As a result, the transmission of a broadband signal experiences a severe SNR penalty and a data rate penalty. Such a scenario can be generally described as a frequency selective slow fading channel or a dispersive channel. A similar effect can also be caused by the low-pass frequency response of the front-end components, such as light emitting diodes (LEDs), photodiodes (PDs), and electronic amplifiers, even in an LOS communication scenario. Therefore, channel equalization at the receiver is required to recover the broadband transmitted signal and to reduce the SNR penalty, which nonetheless still results in an increased SNR requirement to achieve a target BER. However, as shown in Figs. 2.14(a) and 2.18(b), NLOS communication links have similar received signal intensity to LOS communication links at larger distances because of the lower path loss exponent. Furthermore, NLOS communication is more robust to link blockage by a mobile object or a person. Therefore, it is important to determine which digital modulation scheme for optical wireless communication (OWC) is most capable of delivering the highest spectral efficiency for a target electrical SNR for a practical dynamic range of the transmitter, when the bandwidth of the broadband signal exceeds the coherence bandwidth of the channel.

In this chapter, modulation schemes for OWC are studied under the constraints of a practical communication scenario imposed by the optical front-ends and the optical wireless channel. Considering the practical linear dynamic range of the front-end after pre-distortion, an average electrical power constraint and optical power constraints, such as the minimum, average, and maximum optical power constraints, are imposed [43, 133]. In this setup, the spectral efficiency of multi-carrier modulation is compared with the spectral efficiency of single-carrier modulation in the flat fading channel with AWGN [67]. Multi-carrier modulation is represented by optical orthogonal frequency division multiplexing (O-OFDM) with quadrature amplitude modulation (M-QAM). The two general M-QAM O-OFDM schemes for OWC, i.e. direct-current-biased O-OFDM (DCO-OFDM) and asymmetrically clipped O-OFDM (ACO-OFDM), are studied. Single-carrier modulation is realized through pulse modulation techniques, such as multi-level pulse position modulation (M-PPM) and multi-level pulse amplitude modulation (M-PAM). In addition, the achievable information rate of OFDM-based OWC with non-linear clipping distortion due to the limited practical linear dynamic range of the optical front-end is presented for the flat fading channel with AWGN [133]. The study is accommodated within the Shannon framework, in order to show the maximum information rate of O-OFDM, considering the above-mentioned electrical and optical power constraints. Finally, the spectral efficiency of multi-carrier modulation is compared with the spectral efficiency of single-carrier pulse modulation in the dispersive optical wireless channel with AWGN for a practical dynamic range of the optical front-end, when the signal bandwidth significantly exceeds the coherence bandwidth of the channel [67, 134].

5.2   Constraints on the information rate in OWC

The link impairments, such as the non-linear transfer characteristics of the front-ends, the channel characteristics, and the noise sources, are the limiting factors for the system capacity. In a practical system implementation, the computational complexity of the signal processing and the nature of the front-ends, such as the electronic and optical components in an OWC system, impose constraints on the shape of the transmitted signals. In this section, the challenges in the maximization of the capacity of practical OWC system implementations are discussed in the context of the existing literature.

5.2.1   Link impairments

Impairments in the optical wireless link result in signal distortion and capacity degradation. In this study, the signal distortion is caused by the non-linear transfer characteristic of the transmitter, the dispersive optical wireless channel, and the AWGN at the receiver.

Because of the non-negative dynamic range of the transmitter [91], signals with negative portions experience a non-linear distortion due to signal clipping. In addition, imperfections of the optical front-ends due to the use of off-the-shelf components result in a non-linear transfer of the transmitted signal, which introduces further non-linear signal distortion. It is shown in [135] that the non-linear transfer characteristic of the LED can be compensated by pre-distortion. A linear characteristic is obtainable, however, only over a limited range. Therefore, the transmitted signal is constrained between levels of minimum and maximum radiated optical power. Furthermore, the average optical power level is constrained by the eye safety regulations [44] and/or design requirements, such as signal dimming. An example of the latter is OWC in an aircraft cabin [16] where the cabin illumination serves the dual purpose of visible light communication (VLC). In order to condition the signal in accordance with these constraints, signal scaling in the digital signal processor (DSP) and addition of a direct current (DC) bias in the analog circuitry are required.

In OWC with intensity modulation and direct detection (IM/DD), because of the fact that the light carrier wavelength is significantly smaller than the area of the photodetector, there is no fast fading in the channel due to effective spatial diversity [21], but rather only slow fading in the form of path loss and log-normal shadowing [16]. In addition, based on the geometry of the communication link, the optical wireless channel is dominated by LOS or NLOS/multipath signal components. The LOS channel has a very narrow power delay profile with a very broad coherence bandwidth, which enables the transmission of signals with an equally broad information bandwidth and a very high data rate. It can be generally assumed as a frequency non-selective slow fading channel or a flat fading channel. The NLOS channel, however, has a very large RMS delay spread, and therefore a very narrow coherence bandwidth. Thus, it can be modeled as a frequency selective slow fading channel or a dispersive channel. As a result, the transmission of a broadband signal has a data rate penalty. In general, a similar effect is also caused by the low-pass frequency response of the front-end components, such as LEDs, PDs, and electronic amplifiers. However, as presented in Figs. 2.14(a) and 2.18(b), NLOS communication scenarios provide an irradiation intensity similar to LOS communication scenarios at larger distances because of the lower path loss exponent. Furthermore, an NLOS link is significantly more robust to physical-link blockage due to obstruction by a mobile object or person. Therefore, multi-carrier transmission with an inherent robustness to multipath dispersion is expected to deliver very high data rates and high quality of service in medium-range indoor communication scenarios with high mobility.

At the receiver, the signal is detected by a PD and pre-amplified by a transimpedance amplifier (TIA). Here, the ambient light produces shot noise, and thermal noise arises in the electronic circuitry. These noise components are generally white with a Gaussian distribution. As a result, the noise at the receiver can be modeled as AWGN.

5.2.2   On the maximization of information rate

The optical wireless channel is a linear, time-invariant, memoryless channel, where the channel output can be obtained by linear convolution of the impulse response of the channel and the transmitted signal [21]. In general, in OWC systems, the ambient light produces high-intensity shot noise at the receiver. In addition, thermal noise arises due to the electronic pre-amplifier in the receiver front-end. Both of these noise sources can be accurately modeled as AWGN which is independent from the transmitted signal [21]. Therefore, the OWC systems benefit from FEC channel coding procedures, such as turbo codes or low density parity check (LDPC) codes, to approach the Shannon capacity [165] under the average electrical power constraint, where only the alternating current (AC) signal power is considered [167]. In general, in VLC systems the additional DC-bias power that may be required to facilitate a unipolar signal is employed for illumination as a primary functionality. Therefore, it can be excluded from the calculation of the electrical signal power invested in the complementary data communication. In infrared (IR) communication systems, the DC-bias power is constrained by the eye safety regulations [44], and it is generally included in the calculation of the electrical SNR [67]. Therefore, the systems experience an SNR penalty because of the DC bias, and a framework for its minimization is proposed in this book.

The body of literature on the capacity of the band-limited linear optical wireless channel with AWGN mainly differs in the imposition of the constraints on the transmitted signals, e.g. average electrical power constraint, average optical power constraint, and peak optical power constraint. Essiambre *et al.* [167] consider the validity of Shannon capacity [165] as a function of the electrical SNR when only an average electrical power constraint is imposed on the AC electrical power of single-carrier or multi-carrier signals and a linear transfer characteristic of the optical front-end. Hranilovic and Kschischang [168] assume signal non-negativity and an average optical power constraint. They derive an upper bound of the capacity as a function of the optical SNR using the Shannon sphere packing argument [169], and they present a lower bound of the capacity using a maxentropic source distribution. Examples are given for PAM, QAM, and signals in the form of prolate spheroidal waves. Later, Farid and Hranilovic [170, 171] tighten the upper and lower bounds using an exact geometrical representation of signal spaces, and they add a peak optical power constraint. With the increasing popularity of multi-carrier systems, You and Kahn [172] present the capacity of DCO-OFDM using the sphere packing argument [169] under an average optical power constraint, an infinite dynamic range of the transmitter, and a sufficient DC bias to ensure non-negativity. In this case, there is a fixed ratio between the average optical power and the total electrical power, i.e. AC and DC electrical power, as presented in [131]. It is shown that the DCO-OFDM system capacity approaches Shannon capacity [165] at high electrical SNR, leaving a 3-dB gap due to the DC-bias penalty on the SNR given in [131]. Recently, Li *et al.* [173, 174] investigated the information rate of

ACO-OFDM, confirming that it has a very similar form to the Shannon capacity [165]. They also assume an infinite dynamic range of the transmitter and an average optical power constraint. Similarly, there is a fixed ratio between the optical signal power and the electrical signal power, which merely modifies the received electrical SNR, as generalized in [131]. It is shown that ACO-OFDM achieves half of the Shannon capacity due to the half bandwidth utilization, and there is a further 3-dB penalty on the electrical SNR due to the effective halving of the electrical power within the ACO-OFDM framework.

Following the Bussgang decomposition, the mutual information of radio frequency (RF) OFDM systems with non-linear distortion has been studied in [110, 154, 175]. The maximum achievable rate of Gaussian signals with non-linear distortion has been presented in [176], considering the mutual information of the transmitted and received time-domain signals. This information rate can be approached with iterative time-domain signal processing techniques, such as decision-aided signal reconstruction [177] or iterative non-linear noise estimation and cancelation [175, 178], at the expense of an increased computational complexity. In general, practical indoor O-OFDM system implementations aim to reduce the computational effort only to the fast Fourier transform (FFT) operations, while the additional system parameters are computed offline and stored in look-up tables, in order to realize high data rate optical links [9]. In DCO-OFDM and ACO-OFDM, the information-carrying subcarriers are demodulated in the frequency domain, where the non-linear distortion is transformed into additive Gaussian noise. Since the transmitter biasing parameters, such as the signal standard deviation and the DC bias, directly influence the received electrical SNR [131], the optimum biasing setup for given minimum, average, and maximum optical power constraints under an average electrical power constraint is essential for the OWC system information rate.

In this book, the achievable information rates of the DCO-OFDM and ACO-OFDM systems are formulated in the context of non-linear signal distortion at the optical front-end. In addition to the minimum and maximum optical power constraints, an average optical power constraint is imposed as well, and the information rates of the systems are presented, excluding or including the additional DC-bias power in the calculation of the electrical SNR.

5.3   Modulation schemes in the flat fading channel with AWGN

In this section, the multi-carrier DCO-OFDM and ACO-OFDM with M-QAM are compared with the single-carrier M-PPM and M-PAM in terms of the electrical SNR requirement and the spectral efficiency in a flat fading channel. By the use of pre-distortion, the linear dynamic range of the optical front-end can be maximized. The non-linear signal distortion in O-OFDM can be reduced to double-sided signal clipping at normalized bottom and top clipping levels, λ_{bottom} and λ_{top}. Practical linear dynamic ranges of 10 dB, 20 dB, and 30 dB are considered. An average optical power constraint is imposed, and it is varied over the entire dynamic range for the purpose of dimming of the radiated average optical power. The minimum electrical SNR requirement for a target BER of M-QAM O-OFDM is obtained through optimum signal scaling and DC-biasing [43]. Through dimming of the average optical signal power, it is shown that there is still sufficient electrical signal power for communication.

5.3.1 Biasing optimization of Gaussian signals

The choice of the optimum biasing parameters in O-OFDM, such as the signal variance, σ^2, and the DC bias, β_{DC}, when considering front-end optical power constraints, such as the normalized minimum optical power constraint, $P_{min,norm}$, the normalized maximum optical power constraint, $P_{max,norm}$, and the normalized average optical power constraint, $P_{avg,norm}$, for a given QAM modulation order, M, can be formulated as an optimization problem. The objective of the optimization is the minimization of the electrical SNR requirement to achieve a target BER, which is summarized in Table 5.1. Here, the electrical SNR requirement is represented by the electrical SNR per bit, $\gamma_{b(elec)} = E_{b(elec)}/N_0$, i.e. the average electrical bit energy normalized to the power spectral density of the AWGN. It is expressed from (4.12) by the use of the effective received electrical SNR per bit, $\Gamma_{b(elec)}$, from (4.23), where no bit and power loading is

Table 5.1 Minimization of $\gamma_{b(elec)}$ over σ and β_{DC} for given target BER, M, $P_{min,norm}$, $P_{max,norm}$, and $P_{avg,norm}$

Given:
BER, M, $P_{min,norm}$, $P_{max,norm}$, and $P_{avg,norm}$

Find:
$$\operatorname*{argmin}_{\substack{\sigma \geq 0 \\ \beta_{DC} \geq 0}} \gamma_{b(elec)}(\sigma, \beta_{DC}) \geq 0$$
where

> ZF equalizer

$$\gamma_{b(elec)} = \frac{G_B}{|H(f_{info})|^2 G_T G_{DC}} \left(qK^2 - \frac{G_B \log_2(M)\sigma_{clip}^2}{P_{s(elec)}} \right)^{-1}$$

> MMSE equalizer

$$\gamma_{b(elec)} = \frac{\dfrac{G_B}{G_T G_{DC}} - \left(qK^2 - \dfrac{G_B \log_2(M)\sigma_{clip}^2}{P_{s(elec)}} \right)}{|H(f_{info})|^2 \left(qK^2 - \dfrac{G_B \log_2(M)\sigma_{clip}^2}{P_{s(elec)}} \right)}$$

$$q = \frac{3 \log_2(M)}{M-1} \left(Q^{-1} \left(\frac{BER\sqrt{M} \log_2(M)}{4(\sqrt{M}-1)} \right) \right)^{-2}$$

Constraints: $E[\Phi(x_l)] \leq P_{avg,norm}$
$\lambda_{top} > \lambda_{bottom}$ in DCO-OFDM
$\lambda_{top} > \lambda_{bottom} \geq 0$ in ACO-OFDM

utilized. This is because of the considered flat fading channel with impulse response of $h(t) = \delta(t)$, where $\delta(t)$ is the Dirac delta function, and a respective frequency response of $|H(f_{info})|^2 = 1$. In addition, only the modulation orders, M, of the even powers of 2 from Table 4.1 are used in (4.12) for simplicity.

The optimization problem from Table 5.1 has a trivial solution when the DC-bias power is not included in the calculation of the effective electrical SNR per bit, $\Gamma_{b(elec)}$, i.e. when $G_{DC} = 1$. From (3.19), it follows that K^2 decreases when the signal is more severely clipped. In addition, because of the fact that the clipping noise variance is non-negative, $\Gamma_{b(elec)}$ is maximized when the signal clipping is minimized. For instance, such a clipping scenario in DCO-OFDM is represented by $\lambda_{bottom} = -4$ and $\lambda_{top} = 4$. It is similar to the one used in [138], in order to minimize the clipping distortion. The equivalent scenario for ACO-OFDM is $\lambda_{bottom} = 0$ and $\lambda_{top} = 4$. These setups enable modulation orders as high as $M = 1024$ with a deviation from the true minimum required $\gamma_{b(elec)}$ of only 0.1 dB at BER of 10^{-3}.

However, the optimization problem has a non-trivial solution when the DC-bias power is included in the calculation of the effective electrical SNR per bit, $\Gamma_{b(elec)}$, i.e. when $G_{DC} < 1$. The DC-bias gain, G_{DC}, is obtained from (4.20) and (4.21) in DCO-OFDM and ACO-OFDM, respectively. The bandwidth and time utilization factors are denoted in O-OFDM as G_B and G_T. The clipping distortion parameters, such as the attenuation factor, K, and the clipping noise variance, σ_{clip}^2, can be obtained in DCO-OFDM and ACO-OFDM from (3.19), (3.22), and (3.24). Here, $P_{s(elec)} = \sigma^2$ stands for the average electrical symbol power, and $Q(\cdot)$ is the complementary cumulative distribution function (CCDF) of a standard normal distribution with zero-mean and unity variance. The average optical power level of the transmitted O-OFDM signal is denoted as $E[\Phi(x_l)]$. It can be obtained from (4.15), and it is constrained to $P_{avg,norm}$. The objective function is presented for the cases of a zero forcing (ZF) and a minimum mean-squared error (MMSE) equalizer. The analytical approach to solve the minimization problem, such as for example the method of Lagrange multipliers [179], leads to a system of non-linear transcendental equations, which does not have a closed-form solution. Therefore, a numerical optimization procedure is required, and the minimization can be carried out through a computer simulation for a particular choice of front-end optical power constraints. The formal proof of convexity of the objective function from Table 5.1 over the constrained function domain is equally intractable as the analytical minimization approach. However, the convexity can be illustrated by means of a computer simulation. The objective function from Table 5.1 is illustrated in Figs. 5.1 and 5.2 for DCO-OFDM and ACO-OFDM, respectively, in the flat fading channel, for a 10-dB linear dynamic range and a ZF equalizer. It is shown that this objective function has a unique optimum convex region. Therefore, an iterative optimization procedure based on the gradient descent method can be employed to find the optimum solution [179]. An example of the method is illustrated in Fig. 5.3 for a one-dimensional convex function with initial condition within the feasible region. In this study, the constrained optimization function *fmincon* from the optimization toolbox of the computation software MathWorks Matlab [180] is used.

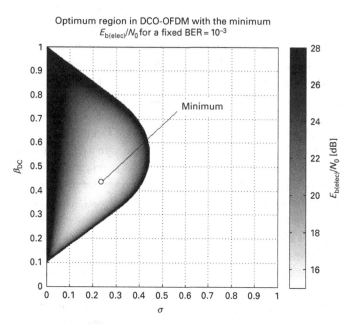

Figure 5.1 Convex objective function of σ and β_{DC} in DCO-OFDM with the minimum $E_{b(elec)}/N_0$ for a BER = 10^{-3} with 4-QAM. $P_{min,norm} = 0.1$ and $P_{max,norm} = 1$ are considered in addition to a flat fading channel with $h(t) = \delta(t)$ and a linear ZF equalizer. The DC-bias power is included in the electrical SNR

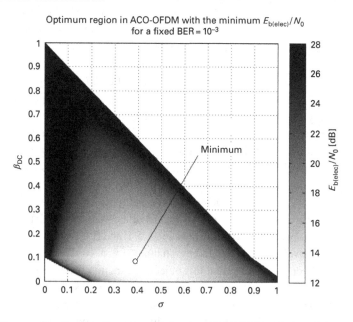

Figure 5.2 Convex objective function of σ and β_{DC} in ACO-OFDM with the minimum $E_{b(elec)}/N_0$ for a BER = 10^{-3} with 4-QAM. $P_{min,norm} = 0.1$ and $P_{max,norm} = 1$ are considered in addition to a flat fading channel with $h(t) = \delta(t)$ and a linear ZF equalizer. The DC-bias power is included in the electrical SNR

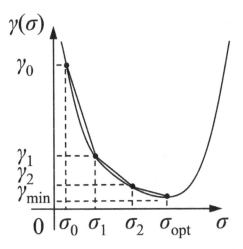

Figure 5.3 Example of an iterative optimization procedure based on the gradient descent method. A one-dimensional convex function, $\gamma(\sigma)$, is displayed. Initial condition, σ_0, is chosen within the feasible region. The optimum solution, $(\sigma_{opt}, \gamma_{min})$, is illustrated

5.3.2   Maximum spectral efficiency without an average optical power constraint

In this study, M-QAM O-OFDM is compared with M-PPM and M-PAM in the flat fading channel without dispersion, i.e. $h(t) = \delta(t)$, in terms of the electrical SNR requirement, $\gamma_{b(elec)}$, and the corresponding spectral efficiency. A target BER of 10^{-3} is considered, since it is the minimum requirement to successfully enable FEC channel coding. First, the DC-bias power is excluded from the calculation of the electrical SNR requirement, i.e. $G_{DC} = 1$ is assumed. The following modulation orders are chosen: $M = \{2, 4, 16, 64, 256, 1024\}$. Here, single-carrier binary phase shift keying (BPSK) is identical to 2-PAM. A 10-dB dynamic range of the optical front-end is assumed. The transmitted signal spans the entire dynamic range of optical power, and no constraint is imposed on the radiated average optical power, in order to obtain the best BER system performance for the given dynamic range. In single-carrier transmission, non-linear signal distortion is avoided through pre-distortion. The average optical power level in M-PPM is set to $E[F(\mathbf{x}_l)] = P_{max,norm}/M$, and in M-PAM to $E[F(\mathbf{x}_l)] = (P_{min,norm} + P_{max,norm})/2$. In multi-carrier transmission, a large number of subcarriers, e.g. 2048, is chosen. Since $G_{DC} = 1$ is assumed, minimum signal clipping is assumed in O-OFDM, i.e. $\lambda_{bottom} = -4$ and $\lambda_{top} = 4$ in DCO-OFDM, and $\lambda_{bottom} = 0$ and $\lambda_{top} = 4$ in ACO-OFDM. These setups enable modulation orders as high as $M = 1024$ with a deviation from the true minimum required $\gamma_{b(elec)}$ of only 0.1 dB at BER of 10^{-3}. In both systems, the average optical power level, $E[\Phi(\mathbf{x}_l)]$, can be obtained from (4.15). The resulting spectral efficiency vs. the electrical SNR requirement plot of the transmission schemes for OWC is presented in Fig. 5.4. It is shown that PPM is the only system which can operate at very low SNR in the range of 4.1–6.8 dB. For a given higher SNR, DCO-OFDM and PAM demonstrate equal highest spectral efficiency.

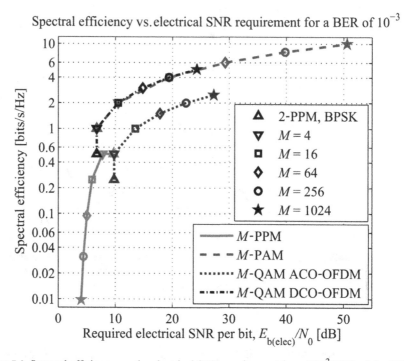

Figure 5.4 Spectral efficiency vs. the electrical SNR requirement for a 10^{-3} BER of the OWC schemes in a flat fading channel with impulse response $h(t) = \delta(t)$ and a neglected DC-bias power [67]. ©2012 IEEE

When the DC-bias power is counted towards the signal power, i.e. $G_{DC} < 1$, the transmission schemes for OWC incur an SNR penalty because the DC bias reduces the useful AC signal power for a fixed total signal power. As a result, the electrical SNR requirement to achieve the target BER of 10^{-3} in the chosen biasing setup is increased. Based on the different signal statistics, the compared systems incur a different SNR penalty due to the DC bias. The DC-bias gain factor, G_{DC}, can be obtained for M-PAM from (4.8), (4.10), and (4.11). In DCO-OFDM and ACO-OFDM, the DC-bias gain is given in (4.20) and (4.21), respectively. The electrical SNR requirement of O-OFDM is minimized by numerically solving the optimization problem from Table 5.1 for a 10-dB dynamic range of the transmitter and QAM modulation orders, $M = \{4, 16, 64, 256, 1024\}$, with linear ZF equalizer in a flat fading channel. The details of the optimum biasing parameters are presented in Table 5.2. Considering the electrical power invested in the DC bias, it is shown that DCO-OFDM requires a non-symmetric clipping setup with the DC bias placed below the middle of the dynamic range, in order to minimize the required $\gamma_{b(elec)}$ for target BER. The optimum performance of ACO-OFDM is obtained when the downside clipping is kept at a minimum by setting the DC bias close to $P_{min,norm}$.

The spectral efficiency *vs.* the electrical SNR requirement plot in the case of included DC-bias power is presented in Fig. 5.5. It is shown that the multi-carrier modulation

Table 5.2 Optimum biasing parameters, σ and β_{DC}, and optimum normalized clipping levels, λ_{bottom} and λ_{top}, in ACO-OFDM and DCO-OFDM for a given M-QAM order. A 10-dB dynamic range is considered in addition to a target BER of 10^{-3}, a flat fading channel with impulse response $h(t) = \delta(t)$, and a linear ZF equalizer. The DC-bias power is included in the electrical SNR.

Modulation	ACO-OFDM				DCO-OFDM			
	σ	β_{DC}	λ_{bottom}	λ_{top}	σ	β_{DC}	λ_{bottom}	λ_{top}
4-QAM	0.39	0.08	0.05	2.38	0.23	0.44	−1.47	2.42
16-QAM	0.33	0.09	0.01	2.76	0.18	0.48	−2.11	2.87
64-QAM	0.29	0.1	0	3.13	0.15	0.5	−2.6	3.27
256-QAM	0.26	0.1	0	3.48	0.14	0.51	−3.03	3.63
1024-QAM	0.24	0.1	0	3.81	0.12	0.52	−3.41	3.97

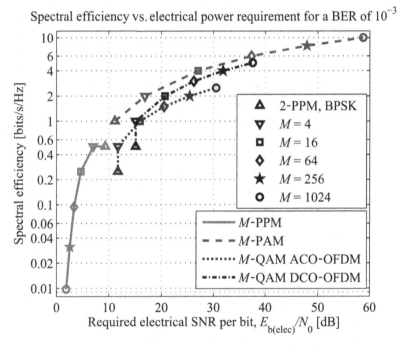

Figure 5.5 Spectral efficiency vs. the electrical SNR requirement for a 10^{-3} BER of the OWC schemes in a flat fading channel with impulse response $h(t) = \delta(t)$, including the DC-bias power for a 10-dB dynamic range

schemes incur a larger DC-bias penalty as compared to single-carrier transmission. For the considered 10-dB dynamic range of the optical front-end, the difference in the electrical SNR requirement between PAM and DCO-OFDM for a target spectral efficiency grows from 4 dB to 5.5 dB for the considered modulation orders. The increase of the gap is due to the fact that higher-order modulation is more vulnerable to non-linear distortion noise. As shown in Table 5.2, the optimum biasing requires higher clipping levels, i.e. a smaller portion of the signal peaks can be clipped, and therefore a higher DC-bias penalty is incurred. DCO-OFDM consistently outperforms ACO-OFDM, and the SNR gap increases for higher spectral efficiencies. PPM remains the best-performing candidate at very low SNR with a 5-dB difference between 4-PPM and 4-QAM ACO-OFDM.

5.3.3   Spectral efficiency with an average optical power constraint

The eye safety regulations [44] and/or design requirements, such as signal dimming capabilities, impose a constraint on the average level of radiated optical power. Dimming is generally achieved either by decreasing the duty cycle of the signal or by decreasing the forward current through the LED. While the former approach finds greater practical adoption due to its simple implementation and a linear relation between duty cycle and output radiance, the latter approach offers greater chromatic stability and higher luminous efficacy [181–183]. PPM is one technique to implement LED dimming through reduction of the duty cycle, and it is suitable for applications with very low optical power output. On the other hand, the OFDM-based techniques, DCO-OFDM and ACO-OFDM, as well as the single-carrier, PAM, can be employed to modify the average level of the forward current through the LED, and these techniques allow for a higher optical power output. Therefore, these modulation schemes are considered for the study with the average optical power constraint.

In this section, first, the O-OFDM techniques are compared in terms of the spectral efficiency and the electrical SNR requirement for an equal average optical power output. Here, the average optical power constraint in Table 5.1 is met with equality, i.e. $E[\Phi(\mathbf{x}_l)] = P_{\text{avg,norm}}$ in order to illustrate the influence of signal dimming the electrical SNR requirement. Linear dynamic ranges of 10 dB, 20 dB, and 30 dB are considered, and, therefore, the normalized minimum optical power is set to $P_{\text{min,norm}} = \{0.1, 0.01, 0.001\}$. In addition, the comparison is extended with the single-carrier PAM scheme, where the average optical power level is set through the DC bias, β_{DC}, as outlined in Section 4.3.2.

When the DC-bias power is not counted towards the electrical SNR, the modulation schemes attain spectral efficiency over the SNR range given in Fig. 5.4 for any average optical power level. When the DC-bias power is included, operation at low and high optical power levels results in a large DC-bias penalty. In the following, this behavior is studied in detail.

For the comparison of the SNR requirement of DCO-OFDM and ACO-OFDM under an average optical power constraint, the following QAM orders are chosen: $M = \{4, 16, 64, 256, 1024\}$. A target BER of 10^{-3} is considered, and a ZF equalizer is used. Plots of the electrical SNR requirement vs. normalized average optical power are presented for DCO-OFDM and ACO-OFDM over linear dynamic ranges of 10 dB, 20 dB, and 30 dB in Figs. 5.6, 5.7, and 5.8, respectively. As illustrated in Fig. 5.7, the small slopes of the graphs in the middle of the dynamic range suggest that average optical power over more than 50% and 25% of the dynamic range can be supported by an SNR margin as low as 3 dB in DCO-OFDM and ACO-OFDM, respectively. It is shown that for an equal modulation order, M, DCO-OFDM outperforms ACO-OFDM

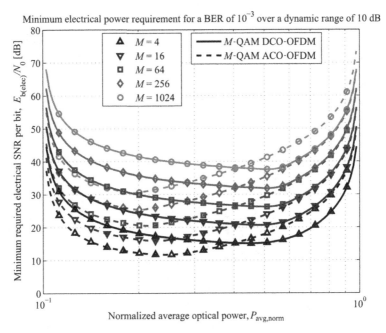

Figure 5.6 Minimum electrical SNR requirement in DCO-OFDM and ACO-OFDM for a 10^{-3} BER vs. average optical power over a 10-dB dynamic range [43]. ©2012 IEEE

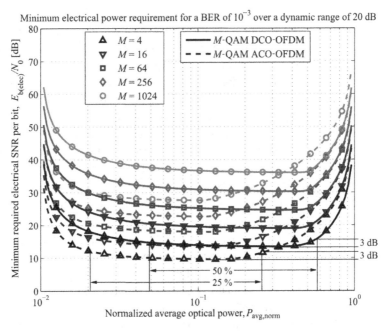

Figure 5.7 Minimum electrical SNR requirement in DCO-OFDM and ACO-OFDM for a 10^{-3} BER vs. average optical power over a 20-dB dynamic range [43]. ©2012 IEEE

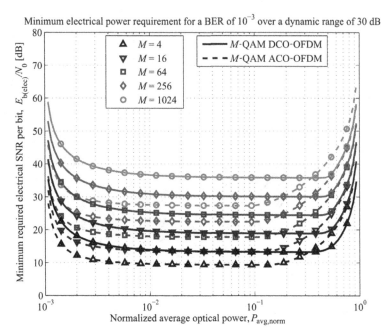

Figure 5.8 Minimum electrical SNR requirement in DCO-OFDM and ACO-OFDM for a 10^{-3} BER vs. average optical power over a 30-dB dynamic range [43]. ©2012 IEEE

in terms of the minimum electrical SNR requirement for average optical power levels in the upper part of the dynamic range, whereas ACO-OFDM prevails for lower average optical power levels because of the respective Gaussian and half-Gaussian distributions of the signals. In addition, the minimum electrical SNR requirement graphs exhibit an absolute minimum. This suggests that there is an average optical power level that allows for the best joint maximization of the signal variance, minimization of the clipping distortion, and minimization of the DC-bias penalty as given in Table 5.1. For both systems, the absolute minimum electrical SNR requirement and the corresponding average optical power level are presented in Fig. 5.9. They decrease with an increase of the dynamic range because of the relaxed $P_{\text{min,norm}}$ and $P_{\text{max,norm}}$ constraints in the optimization. The details of the optimum biasing parameters for the 10-dB dynamic range are given in Table 5.2.

In order to find out which O-OFDM system delivers the higher throughput for a given average optical power level, the spectral efficiency and the minimum electrical SNR requirement are shown in Fig. 5.10 for an average optical power level dimmed down to 20%. DCO-OFDM is expected to have a lower electrical SNR requirement and superior spectral efficiency as compared to ACO-OFDM for average optical power levels in the upper part of the dynamic range. However, this is shown to be the case also for low average optical power levels, i.e. $P_{\text{avg,norm}} = 0.2$, as the dynamic range increases. Here, LEDs with wider linear dynamic ranges are proven to be the enabling factors for OWC with low optical power radiation. In addition, DCO-OFDM demonstrates a lower

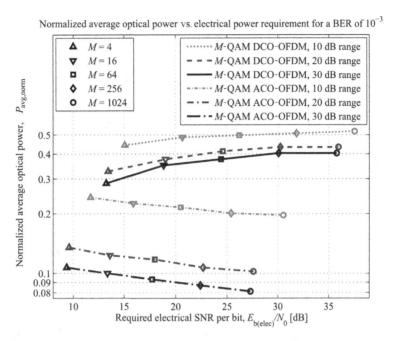

Figure 5.9 Normalized average optical power vs. absolute minimum electrical SNR requirement in DCO-OFDM and ACO-OFDM for a 10^{-3} BER [43]. ©2012 IEEE

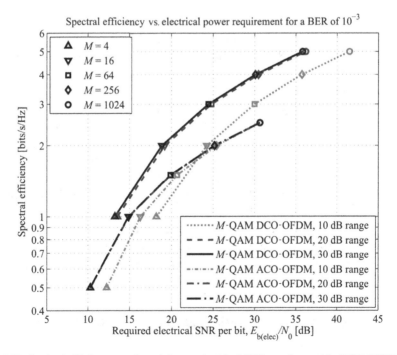

Figure 5.10 Spectral efficiency vs. the minimum electrical SNR requirement in DCO-OFDM and ACO-OFDM for a 10^{-3} BER and an average optical power level set to 20% of the maximum of the dynamic range [43]. ©2012 IEEE

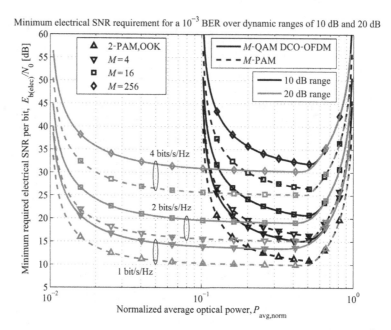

Figure 5.11 Minimum electrical SNR requirement in DCO-OFDM and PAM for a 10^{-3} BER vs. average optical power over dynamic ranges of 10 dB and 20 dB

minimum electrical SNR requirement for a target BER as compared to ACO-OFDM for modulation orders with similar spectral efficiencies for average optical power levels over more than 85%, 90%, and 95% of the dynamic ranges of 10 dB, 20 dB, and 30 dB, respectively.

The comparison is extended with the performance of the single-carrier PAM scheme under an average optical power constraint. Here, DCO-OFDM is compared with PAM for dynamic ranges of 10 dB and 20 dB. Modulation orders with spectral efficiencies of 1, 2, and 4 bits/s/Hz are chosen. The electrical SNR requirement vs. normalized average optical power plot is presented in Fig. 5.11. In accordance with the performance over the 10-dB dynamic range from Fig. 5.5, PAM shows a lower SNR requirement as compared to DCO-OFDM, and it ranges from 4 dB to 5.5 dB with the increase of spectral efficiency. When a larger linear dynamic range of 20 dB is considered, the gap between DCO-OFDM and PAM is reduced to 3 dB at 1 bit/s/Hz, and to 5 dB at 4 bits/s/Hz. This is because of the relaxed minimum optical power constraint, $P_{min,norm}$, in the optimization of the biasing parameters from Table 5.1.

5.4   Information rate of OFDM-based modulation with non-linear distortion

In this section, the information rate of the multi-carrier O-OFDM schemes, DCO-OFDM and ACO-OFDM, is presented based on the mutual information between transmitted symbols and received symbols in the flat fading channel for a practical linear dynamic range of the transmitter. In practice, the information rate of these modulation schemes can be achieved by means of FEC. Through pre-distortion, the non-linear distortion of the transmitter can be reduced to double-sided signal clipping. Since the clipping distortion merely results into an electrical SNR penalty, the information rate of O-OFDM with non-linear distortion can be accommodated within the Shannon framework. In addition, the average optical power constraint is imposed, and the information rate of O-OFDM with signal dimming is presented, including or excluding the additional DC-bias power in the calculation of the electrical SNR [133].

5.4.1   Biasing optimization of Gaussian signals

Passing through the transmitter front-end, the information-carrying O-OFDM signal, x_l, is subjected to the non-linear distortion function $\Xi(x)$. Through pre-distortion, non-linearity is reduced to double-sided signal clipping, $\Phi(x)$. As a result, the linear dynamic range of the optical front-end is maximized for an optimum achievable information rate. Because of the Gaussian signal distribution in O-OFDM, the Bussgang theorem [137] can be applied, and the non-linear distortion can be modeled as an attenuation of the signal power and introduction of uncorrelated zero-mean non-Gaussian noise. After passing through the FFT at the receiver, the orthogonality of the attenuated information-carrying subcarriers is preserved, and the non-linear distortion noise is transformed into complex-valued Gaussian noise according to the central limit theorem (CLT) [136]. The generalized model for the non-linear distortion of the received information-carrying subcarriers in O-OFDM is illustrated in Fig. 4.8. As a result, the non-linear distortion can be modeled as the transformation of the electrical SNR on an enabled subcarrier presented in (5.2). Because of Hermitian symmetry within the OFDM frame, the DCO-OFDM and ACO-OFDM systems enable $G_B N/2$ orthogonal complex-valued channels, the equivalent of $G_B N$ orthogonal real-valued channels relevant for OWC. The QAM-modulated information-carrying symbols on the orthogonal subcarriers, f_{info}, have real and imaginary parts with uniform distributions. By means of symbol shaping and FEC coding such signals have also been shown to approach the Shannon capacity [166]. Therefore, the two OFDM-based OWC systems can be accommodated within the Shannon framework [165], taking into account the non-linear distortion in any given front-end biasing setup. The mutual information, C, in bits per real dimension (bits/dim) [99] can be expressed in O-OFDM as a function of the undistorted electrical SNR per bit, $\gamma_{b(elec)} = E_{b(elec)}/N_0$, as follows:

$$C = \frac{R_b}{B} = G_T \frac{G_B}{2} \log_2 \left(1 + \text{SNR}\right)$$

$$= G_T \frac{G_B}{2} \log_2 \left(1 + \frac{2K^2 C/G_B}{\dfrac{2C\sigma_{clip}^2}{P_{s(elec)}} + \dfrac{G_B \gamma_{b(elec)}^{-1}}{|H(f_{info})|^2 G_T G_{DC}}}\right). \qquad (5.1)$$

Table 5.3 Minimization of $\gamma_{b(elec)}$ over σ and β_{DC} for given C, $P_{min,norm}$, $P_{max,norm}$, and $P_{avg,norm}$

Given: $C, P_{min,norm}, P_{max,norm},$ and $P_{avg,norm}$

Find: $\underset{\substack{\sigma \geq 0 \\ \beta_{DC} \geq 0}}{\text{argmin}} \; \gamma_{b(elec)}(\sigma, \beta_{DC}) \geq 0$

where

$$\gamma_{b(elec)} = \frac{G_B^2}{2|H(f_{info})|^2 G_T G_{DC} C} \left(\frac{K^2}{2^{2C/(G_T G_B)} - 1} - \frac{G_B \sigma_{clip}^2}{P_{s(elec)}} \right)^{-1}$$

Constraints: $E[\Phi(x_l)] \leq P_{avg,norm}$

$\lambda_{top} > \lambda_{bottom}$ in DCO-OFDM

$\lambda_{top} > \lambda_{bottom} \geq 0$ in ACO-OFDM

The effective electrical SNR on an enabled subcarrier in O-OFDM is given as follows:

$$\text{SNR} = \frac{K^2 P_{s(elec)}/G_B}{\sigma_{clip}^2 + \dfrac{G_B \sigma_{AWGN}^2}{|H(f_{info})|^2 G_{DC}}}, \tag{5.2}$$

where the variance of the AWGN is denoted as σ_{AWGN}^2. Given a limited linear dynamic range between $P_{min,norm}$ and $P_{max,norm}$, the optimum choice of the biasing parameters, such as the signal variance, σ^2, and the DC bias, β_{DC}, can be formulated as an optimization problem. The objective of the optimization is the minimization of the electrical SNR requirement, $\gamma_{b(elec)}$, to achieve a target information rate, C, for a given average optical power constraint, $P_{avg,norm}$. The minimum SNR requirement and the maximum information rate are achieved in a flat fuding channel with an impulse response of $h(t) = \delta(t)$ and a respective frequency response of $|H(f_{info})|^2 = 1$. This optimization problem is summarized in Table 5.3. First, the DC-bias power is neglected in the calculation of the effective electrical SNR in (5.2), i.e. G_{DC} is considered as unity. From (3.19), it follows that K^2 decreases when the signal is more severely clipped. In addition, the clipping noise variance is non-negative as confirmed in (3.22) and (3.24). Therefore, the effective electrical SNR and the information rate are maximized when the signal clipping is minimized. As a result, the optimum biasing setup is achieved by setting the normalized clipping levels, λ_{bottom} and λ_{top}, further apart as the information rate requirement increases, in order to accommodate a larger portion of the signal peaks [67].

A non-trivial optimization problem arises when the DC-bias power is included in the calculation of the effective electrical SNR, i.e. $G_{DC} < 1$ is considered. The analytical approach to solve the minimization problem leads to a system of non-linear transcendental equations which does not have a closed-form solution. Therefore, a numerical

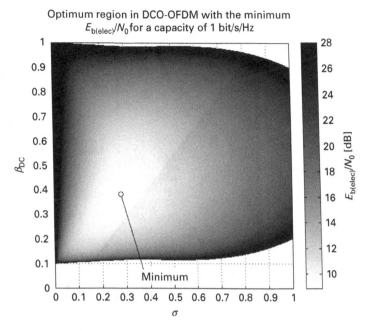

Figure 5.12 Convex objective function of σ and β_{DC} in DCO-OFDM with the minimum $E_{b(elec)}/N_0$ for an information rate of 1 bit/dim [133]. $P_{min,norm} = 0.1$ and $P_{max,norm} = 1$ are considered. The DC-bias power is included in the electrical SNR. ©2013 IEEE

optimization procedure is required, and the minimization can be carried out through computer simulation. In general, the formal proof of convexity of the objective function from Table 5.3 over the constrained function domain is equally intractable as the analytical minimization approach. However, the convexity can be illustrated by means of a computer simulation in Figs. 5.12 and 5.13 for DCO-OFDM and ACO-OFDM, respectively. According to the Bussgang decomposition, the signal and the non-linear distortion noise are uncorrelated, but dependent. Therefore, the information rates obtained by solving the optimization problem from Table 5.3 for $G_{DC} < 1$ and a given set of constraints can be considered as a lower bound on the capacity of the O-OFDM systems.

5.4.2 Maximum information rate without an average optical power constraint

The transmitter front-end constrains $P_{min,norm}$ and $P_{max,norm}$, while $P_{avg,norm}$ is independently imposed by the eye safety regulations and/or the design requirements. In general, constraining the average optical power level to $E[\Phi(\mathbf{x}_l)] \leq P_{avg,norm}$ results in a suboptimum SNR requirement for a target information rate. The minimum SNR requirement is obtained when this constraint is relaxed, i.e. when $E[\Phi(\mathbf{x}_l)]$ is allowed to assume any level in the dynamic range between $P_{min,norm}$ and $P_{max,norm}$. The optimized signal biasing setup is compared with a setup where the suboptimum signal scaling and DC-biasing lead to considerable signal clipping distortion and DC-bias penalty.

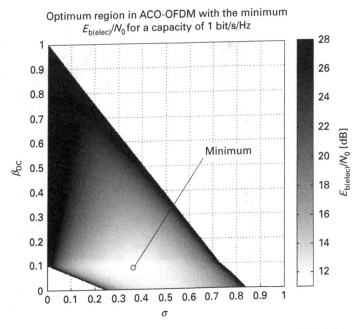

Figure 5.13 Convex objective function of σ and β_{DC} in ACO-OFDM with the minimum $E_{b(elec)}/N_0$ for an information rate of 1 bit/dim [133]. $P_{min,norm} = 0.1$ and $P_{max,norm} = 1$ are considered. The DC-bias power is included in the electrical SNR. ©2013 IEEE

In DCO-OFDM, such a suboptimum setup is realized, for instance when $\beta_{DC} = 0.55$ and $\sigma = 0.2$. This results in normalized clipping levels of $\lambda_{top} = -\lambda_{bottom} = 2.2$ for a 10-dB dynamic range. In ACO-OFDM, the suboptimum biasing parameters are chosen as follows: $\beta_{DC} = 0.2$ and $\sigma = 0.4$, which, respectively yield $\lambda_{bottom} = 0$ and $\lambda_{top} = 2$. The information rate for a 10-dB dynamic range of the optical front-end without an average optical power constraint is presented in Fig. 5.14. When the DC-bias power is not counted towards the signal power, the optimized DCO-OFDM system achieves Shannon capacity. Because of the effective halving of the electrical signal power and the half bandwidth utilization, the optimized ACO-OFDM system exhibits a 3-dB gap to the capacity, which increases with the increase of the target information rate as presented in [51, 173]. In addition, it is shown that the severe signal clipping in the O-OFDM systems without optimization of the biasing setup introduces a negligible SNR penalty at low information rate targets, where the AWGN is dominant. The SNR penalty increases with the increase of the information rate target, where the clipping noise is dominant. When the DC-bias power is added to the signal power, the systems incur an SNR penalty, because the DC bias reduces the useful AC signal power for a fixed total signal power. The optimized DCO-OFDM and ACO-OFDM systems exhibit a gap in the Shannon capacity of 5.6 dB and 5.3 dB, respectively, at 0.1 bits/dim, and 7.1 dB and 9.4 dB at 1 bit/dim. ACO-OFDM has a slightly lower SNR requirement as compared to DCO-OFDM at low information rate targets, and a significantly higher SNR requirement at high information rate targets. It is shown that the optimization of

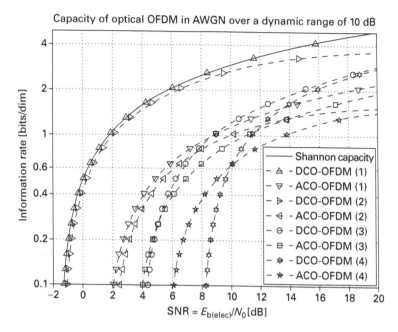

Figure 5.14 Mutual information in DCO-OFDM and ACO-OFDM vs. the electrical SNR requirement for a 10-dB dynamic range without an average optical power constraint: (1) with optimization, DC-bias power not included; (2) without optimization, DC-bias power not included; (3) with optimization, DC-bias power included; (4) without optimization, DC-bias power included [133]. ©2013 IEEE

the biasing setup can reduce the SNR penalty significantly. In this considered scenario, a reduction of 4 dB and 2 dB is observed for DCO-OFDM and ACO-OFDM, respectively, at 0.1 bits/dim, and 2.5 dB at 1 bit/dim. With the increase of the of the linear dynamic range of the optical front-end to 20 dB presented in Fig. 5.15, the clipping distortion is reduced, and the O-OFDM without optimization has a slight increase in information rate. The optimized O-OFDM systems preserve their maximum information rate in the case where the DC-bias power is not included in the calculation of the electrical SNR. The increase of the dynamic range of the optical front-end further reduces the SNR requirement of the optimized systems when the DC-bias power is counted towards the signal power. In DCO-OFDM and ACO-OFDM, the gap is reduced to 4.4 dB and 3.6 dB, respectively, at 0.1 bits/dim, and to 5.6 dB and 7.4 dB at 1 bit/dim.

5.4.3 Information rate with an average optical power constraint

In the next set of results, the average optical power constraint is imposed with equality, and optimization is employed. DCO-OFDM is expected to have a lower electrical SNR requirement and a superior information rate as compared to ACO-OFDM for average optical power levels in the upper part of the dynamic range, while ACO-OFDM is expected to show a superior performance for average optical power levels in the lower

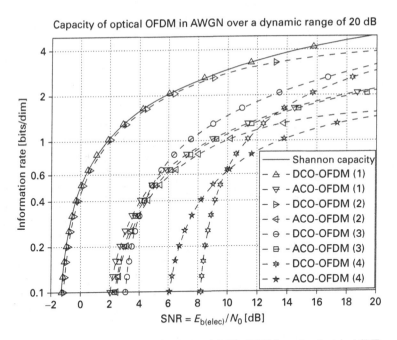

Figure 5.15 Mutual information in DCO-OFDM and ACO-OFDM vs. the electrical SNR requirement for a 20-dB dynamic range without an average optical power constraint: (1) with optimization, DC-bias power not included; (2) without optimization, DC-bias power not included; (3) with optimization, DC-bias power included; (4) without optimization, DC-bias power included [133]. ©2013 IEEE

part of the dynamic range. Therefore, optical power levels of 20% and 50% are chosen for comparison of the optimized O-OFDM systems, and the results for a 10-dB linear dynamic range are presented in Fig. 5.16. When the DC-bias power is not counted towards signal power, both systems achieve their maximum information rate for both average optical power levels, i.e. DCO-OFDM achieves Shannon capacity, while ACO-OFDM has a 3-dB penalty which increases for higher information rate targets. When the DC-bias power is included in the calculation of the electrical SNR, DCO-OFDM completely outperforms ACO-OFDM for the 50% average optical power level. For the 20% average optical power level, ACO-OFDM has a superior information rate as compared to DCO-OFDM only up to the cross-over point of 9.8 dB at 0.8 bits/dim. When the linear dynamic range is increased to 20 dB and the DC-bias power is counted towards the signal power, Fig. 5.17 shows that this cross-over point is shifted towards the lower SNR region, and DCO-OFDM has a superior information rate from 0.3 bits/dim at 4 dB onwards. While the increase of the dynamic range significantly increases the information rate for lower average optical power levels in an optimized biasing setup, the information rate of the higher optical power levels is only negligibly improved. This is because the increase of the dynamic range for a given average optical power level reduces the bottom-level clipping, which is already kept at the minimum for high average optical power levels. For the 20% average optical power level in the increased

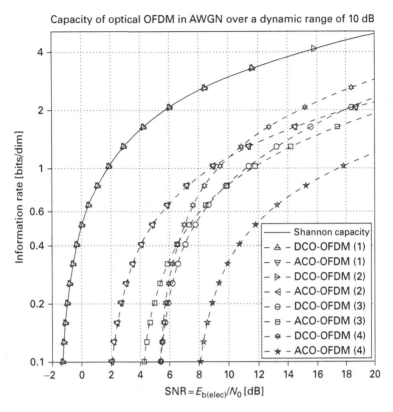

Figure 5.16 Mutual information in DCO-OFDM and ACO-OFDM vs. the electrical SNR requirement for a 10-dB dynamic range with optimization: (1) $P_{\text{avg,norm}} = 0.2$, DC-bias power not included; (2) $P_{\text{avg,norm}} = 0.5$, DC-bias power not included; (3) $P_{\text{avg,norm}} = 0.2$, DC-bias power included; (4) $P_{\text{avg,norm}} = 0.5$, DC-bias power included [133]. ©2013 IEEE

dynamic range of 20 dB, DCO-OFDM and ACO-OFDM exhibit reduction of the SNR penalty of 2.3 dB and 1.7 dB, respectively, at 0.1 bits/dim, and 3.8 dB and 1.7 dB at 1 bit/dim. For the 50% average optical power level, these values amount to merely 0.4 dB and 0.1 dB, respectively, at 0.1 bits/dim, and 0.6 dB and 0.1 dB at 1 bit/dim. When the DC-bias power is excluded from the calculation of the electrical SNR, the optimized O-OFDM systems achieve their maximum information rate also for the increased 20-dB dynamic range.

In order to find out which O-OFDM system delivers the higher information rate for any given average optical power level, the average optical power constraint is swept over the entire linear dynamic range. The solution of the optimization problem from Table 5.3 can be used to iteratively solve the dual problem, i.e. the maximization of the information rate, C, for a target SNR, $\gamma_{\text{b(elec)}}$, and a given average optical power constraint. The information rate of the optimized O-OFDM systems in this scenario is presented in Fig. 5.18 and in Fig. 5.19 for dynamic ranges 10 dB and 20 dB, respectively. Here, SNR targets of $\gamma_{\text{b(elec)}} = 10$ dB and $\gamma_{\text{b(elec)}} = 15$ dB are chosen. When the DC-bias power is not counted towards the signal power, the optimized O-OFDM systems

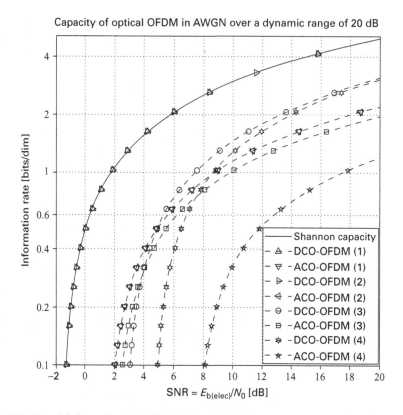

Figure 5.17 Mutual information in DCO-OFDM and ACO-OFDM vs. the electrical SNR requirement for a 20-dB dynamic range with optimization: (1) $P_{avg,norm} = 0.2$, DC-bias power not included; (2) $P_{avg,norm} = 0.5$, DC-bias power not included; (3) $P_{avg,norm} = 0.2$, DC-bias power included; (4) $P_{avg,norm} = 0.5$, DC-bias power included [133]. ©2013 IEEE

consistently achieve their maximum information rate for average optical powers over the entire dynamic range, where DCO-OFDM delivers the higher information rate. When the DC-bias power is included in the calculation of the electrical SNR, DCO-OFDM is shown to have a superior information rate as compared to ACO-OFDM for average optical power levels in the upper part of the dynamic range, while ACO-OFDM shows better performance for lower average optical power levels. This is because of the respective Gaussian and half-Gaussian distributions of the signals. However, as the dynamic range or the target SNR increases, DCO-OFDM is shown to dominate ACO-OFDM over a major part of the lower average optical power levels. Here, DCO-OFDM demonstrates a higher information rate as compared to ACO-OFDM for average optical power levels over more than 89% and 96% of the 10-dB dynamic range for the SNR targets of 10 dB and 15 dB, respectively, and over 99% of the 20-dB dynamic range. In addition, the information rate graphs exhibit an absolute maximum. This suggests that there is an average optical power level which allows for the best joint maximization of the signal variance, minimization of the clipping distortion, and minimization of the DC-bias penalty from Table 5.3. As illustrated in Fig. 5.19, the small

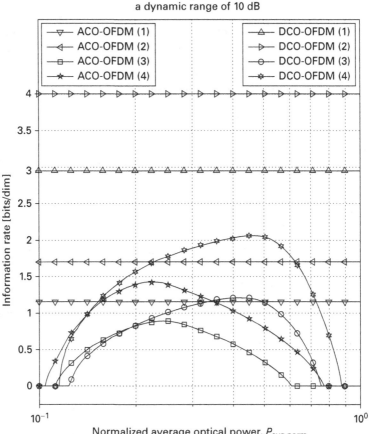

Figure 5.18 Mutual information in DCO-OFDM and ACO-OFDM vs. normalized average optical power for a 10-dB dynamic range with optimization: (1) $E_{b(elec)}/N_0 = 10$ dB, DC-bias power not included; (2) $E_{b(elec)}/N_0 = 15$ dB, DC-bias power not included; (3) $E_{b(elec)}/N_0 = 10$ dB, DC-bias power included; (4) $E_{b(elec)}/N_0 = 15$ dB, DC-bias power included [133]. ©2013 IEEE

gradiant slopes of the graphs in the middle of the dynamic range suggest that average optical powers over more than 50% and 25% of the dynamic range can be supported at the expense of a mere 10% decrease of the information rate in DCO-OFDM and ACO-OFDM, respectively. Therefore, LEDs with wider linear dynamic ranges are proven to be the enabling factor for OWC with low optical power radiation in the case when the DC-bias power is counted towards the signal power. It is important to note that the DC-bias penalty on the electrical SNR, G_{DC}, discussed in [67], is minimized in the DCO-OFDM system as the dynamic range increases. In ACO-OFDM, the DC-bias penalty almost goes to zero as shown in Figs. 5.18 and 5.19. As shown in Section 5.3.3, the point of diminishing returns on the size of the dynamic range appears to be around 20 dB.

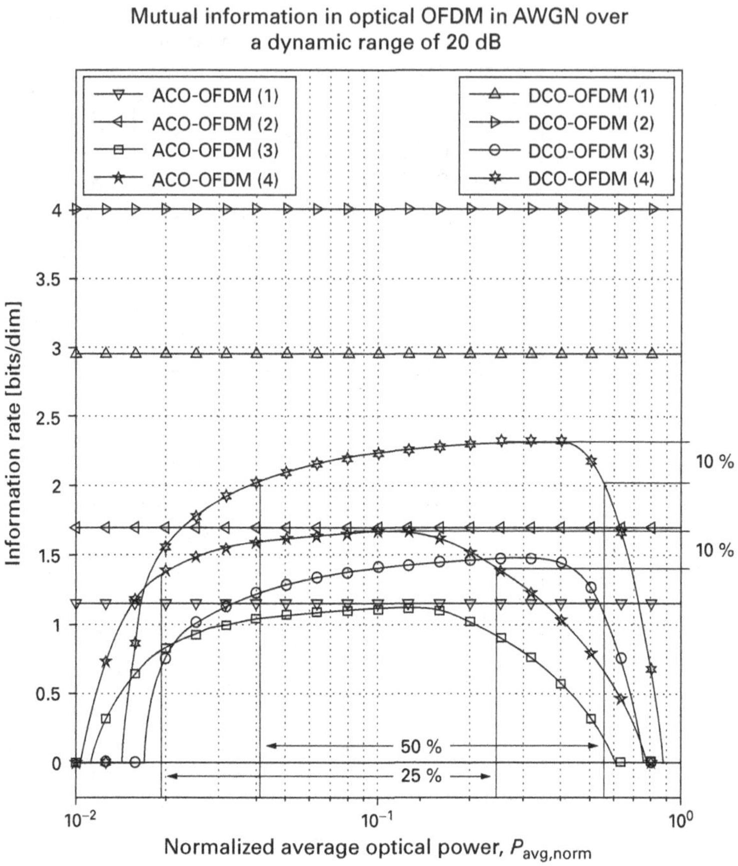

Figure 5.19 Mutual information in DCO-OFDM and ACO-OFDM vs. normalized average optical power for a 20-dB dynamic range with optimization: (1) $E_{b(elec)}/N_0 = 10$ dB, DC-bias power not included; (2) $E_{b(elec)}/N_0 = 15$ dB, DC-bias power not included; (3) $E_{b(elec)}/N_0 = 10$ dB, DC-bias power included; (4) $E_{b(elec)}/N_0 = 15$ dB, DC-bias power included [133]. ©2013 IEEE

5.5   Modulation schemes in the dispersive channel with AWGN

In this section, M-QAM O-OFDM is compared with M-PPM and M-PAM also in the dispersive channel in terms of spectral efficiency, including or excluding the additional DC-bias power in the calculation of the electrical SNR. Through optimum signal scaling and DC-biasing, the O-OFDM signal is conditioned within a 10-dB linear dynamic range. The spectral efficiency of the multi-carrier M-QAM O-OFDM with bit and power loading is compared with the spectral efficiency of the single-carrier M-PPM and M-PAM with linear and non-linear equalization, when the signal bandwidth, B, exceeds the channel coherence bandwidth, B_c [67].

Table 5.4 Minimization of BER over σ and β_{DC} for given target $\gamma_{b(elec)}$, M, $P_{min,norm}$, $P_{max,norm}$, and $P_{avg,norm}$

Given: $\gamma_{b(elec)}$, M, $P_{min,norm}$, $P_{max,norm}$, and $P_{avg,norm}$

Find: $\underset{\substack{\sigma \geq 0 \\ \beta_{DC} \geq 0}}{\mathrm{argmin}} \ \mathrm{BER}(\sigma, \beta_{DC}) \geq 0$

Constraints: $\mathrm{E}\left[\Phi(\mathbf{x}_l)\right] \leq P_{avg,norm}$
  $\lambda_{top} > \lambda_{bottom}$ in DCO-OFDM
  $\lambda_{top} > \lambda_{bottom} \geq 0$ in ACO-OFDM

5.5.1 Biasing optimization of Gaussian signals

The choice of the optimum biasing parameters in O-OFDM, such as the signal variance, σ^2, and the DC bias, β_{DC}, which minimize the average BER over the OFDM frame for a target $\gamma_{b(elec)}$ in the dispersive channel, can be formulated as an optimization problem. For a given ratio between the channel coherence bandwidth and the signal bandwidth, B_c/B, the optimum bit and power loading profile over the OFDM frame that maximizes the average data rate can be obtained by means of the Levin–Campello algorithm [49, 50]. Additional input parameters for the optimization are the front-end optical power constraints, $P_{min,norm}$, $P_{max,norm}$, and $P_{avg,norm}$, and the desired average bit rate per unit bandwidth, equivalent to a QAM modulation order, M. This optimization problem is summarized in Table 5.4. The analytical expression for the average BER can be obtained as an arithmetic average of the subcarrier BER from (4.12), taking into account the bit and power loading profile of the OFDM frame. The additional parameters are taken from Table 4.1 according to the number of loaded bits, i.e. equivalent modulation order. The effective received electrical SNR per bit, $\Gamma_{b(elec)}$, is computed for every subcarrier in the OFDM frame according to (4.23), which is a function of $\gamma_{b(elec)}$. The solution of this optimization problem can be used to iteratively solve the dual problem, i.e. the minimization of the electrical SNR requirement, $\gamma_{b(elec)}$, for a target average BER.

The optimum solution to the optimization problem from Table 5.4 can be readily obtained in the case when the DC-bias power is not included in the calculation of the effective electrical SNR per bit, $\Gamma_{b(elec)}$, i.e. when $G_{DC} = 1$ is considered. From (3.19), it follows that K^2 decreases when the signal is more severely clipped. In addition, because of the fact that the clipping noise variance is non-negative, $\Gamma_{b(elec)}$ is maximized and BER is minimized when the signal clipping is minimized. For instance, such a clipping scenario in DCO-OFDM is represented by $\lambda_{bottom} = -4$ and $\lambda_{top} = 4$. It is similar to the one used in [138], in order to minimize the clipping distortion. The equivalent scenario for ACO-OFDM is $\lambda_{bottom} = 0$ and $\lambda_{top} = 4$. These setups enable modulation orders as high as $M = 1024$ with a deviation from the true minimum required $\gamma_{b(elec)}$ of only 0.1 dB at BER of 10^{-3}.

Including the DC-bias power in the calculation of the effective electrical SNR per bit, $\Gamma_{b(elec)}$, results in a DC-bias penalty, i.e. $G_{DC} < 1$. In this case, the analytical approach to solve the minimization problem leads to a system of non-linear transcendental equations which does not have a closed-form solution. Therefore, a numerical optimization procedure is required, and the minimization can be carried out through a computer simulation for a particular choice of front-end optical power constraints. In general, the formal proof of convexity of the objective function from Table 5.4 over the constrained function domain is equally intractable as the analytical minimization approach. However, the convexity can be illustrated by means of computer simulation. A practical linear dynamic range of 10 dB is assumed. In addition, the average optical power constraint is relaxed in order to obtain the best BER system performance for the given dynamic range of the front-end. The objective function from Table 5.4 is illustrated in Figs. 5.20 and 5.21 for DCO-OFDM and ACO-OFDM, respectively, in the case of a flat fading channel with impulse response of $h(t) = \delta(t)$. It is shown that this objective function has a unique optimum convex region. In OFDM systems, the dispersive channel is represented by a superposition of orthogonal flat fading channels. Therefore, the objective average BER function can be obtained as the average of the BER functions for each flat fading channel which are shown to be convex. Since the expectation operator is a non-negative weighted summation, it preserves the convexity [179]. Therefore, the objective BER function in the dispersive channel remains convex.

5.5.2   DC-bias penalty

In this study, M-QAM O-OFDM is compared with M-PPM and M-PAM also in the dispersive channel in terms of the electrical SNR requirement, $\gamma_{b(\text{elec})}$, to achieve a

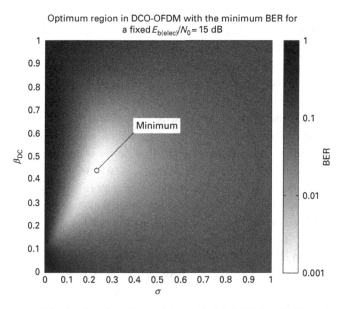

Figure 5.20 Convex objective function of σ and β_{DC} in DCO-OFDM with the minimum BER for a fixed $E_{b(\text{elec})}/N_0 = 15$ dB with 4-QAM [67]. $P_{\text{min,norm}} = 0.1$ and $P_{\text{max,norm}} = 1$ are considered in addition to a flat fading channel with $h(t) = \delta(t)$ and a linear ZF equalizer. The DC-bias power is included in the electrical SNR. ©2012 IEEE

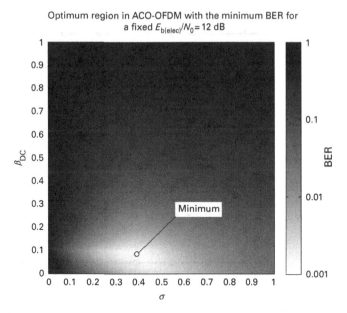

Figure 5.21 Convex objective function of σ and β_{DC} in ACO-OFDM with the minimum BER for a fixed $E_{b(elec)}/N_0 = 12$ dB with 4-QAM [67]. $P_{min,norm} = 0.1$ and $P_{max,norm} = 1$ are considered in addition to a flat fading channel with $h(t) = \delta(t)$ and a linear ZF equalizer. The DC-bias power is included in the electrical SNR. ©2012 IEEE

target BER of 10^{-3} and the corresponding spectral efficiency. When the DC-bias power is included in the calculation of the electrical SNR requirement, the OWC schemes incur an SNR penalty. The DC-bias gain factor, G_{DC}, can be obtained for M-PAM from (4.10), and it is independent of the channel. In order to maximize the average data rate and to minimize the electrical SNR requirement of O-OFDM in the dispersive channel, bit and power loading [49, 50] is combined with the optimum signal biasing from Table 5.4. In DCO-OFDM and ACO-OFDM, the DC-bias gain is given in (4.20) and (4.21), respectively, and it is dependent on the channel because of the employed bit and power loading with optimum signal biasing. Therefore, the resulting optimum σ and β_{DC} are used in the calculation of G_{DC}. The DC-bias gain in the different modulation formats for OWC is presented in Fig. 5.22 as the signal bandwidth exceeds the channel coherence bandwidth. For the considered 10-dB dynamic range, the optimum signal clipping reduces the SNR penalty by up to 6.5 dB for DCO-OFDM and up to 1.4 dB for ACO-OFDM as compared to the minimum signal clipping setup from Section 5.3, where $\lambda_{bottom} = -4$ and $\lambda_{top} = 4$ in DCO-OFDM , and $\lambda_{bottom} = 0$ and $\lambda_{top} = 4$ in ACO-OFDM. In addition, bit and power loading in combination with optimum signal clipping allow the DC-bias gain to saturate above the DC-bias gain in the minimum clipping case. Nevertheless, because of the close to Gaussian and half-Gaussian distribution of the signals, respectively, DCO-OFDM and ACO-OFDM still incur a larger SNR penalty as compared to PAM and PPM, respectively, which have distributions with finite support.

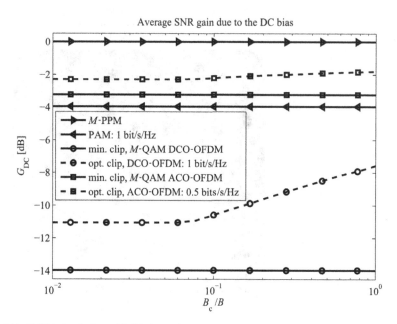

Figure 5.22 DC-bias gain for a 10-dB dynamic range when the signal bandwidth exceeds the channel coherence bandwidth [67]. ©2012 IEEE

5.5.3 Equalizer penalty

In a practical non-flat channel with dispersion [21], the signal bandwidth becomes larger than the channel coherence bandwidth at high data rates. In such a scenario, single-carrier transmission experiences a severe ISI, and it commonly employs a linear feed-forward equalizer (FFE) or a non-linear decision feedback equalizer (DFE). In multi-carrier transmission, a cyclic prefix (CP) is employed which completely eliminates the inter-symbol interference (ISI) and the inter-carrier interference (ICI). In addition, the CP has a negligible impact on the spectral efficiency and the electrical SNR requirement [108]. It transforms the channel into a flat fading channel over the subcarrier bandwidth, and therefore single-tap equalization with bit and power loading [49, 50] can be performed in order to minimize the channel effect. As a result, the equalization process incurs an SNR penalty defined by the equalizer gain factor, G_{EQ}. It can be obtained for single-carrier modulation from (2.19), (2.21), (2.22), or (2.23). In multi-carrier transmission, G_{EQ} is obtained numerically from the average SNR penalty to achieve a target average rate over the OFDM frame for the optimum bit and power loading profile in the dispersive channel. Here, an electrical path gain of $g_{h(elec)} = 1$ is considered for simplicity. The equalizer gain of multi-carrier transmission with bit and power loading and single-tap ZF equalizer is compared with the equalizer gain of single-carrier transmission with multi-tap ZF FFE and ZF DFE as the signal bandwidth grows larger than the channel coherence bandwidth. The result is presented in Fig. 5.23. It is shown that multi-carrier transmission incurs a significantly lower SNR penalty in the equalization process. The plot essentially

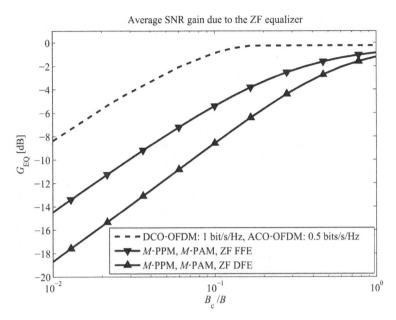

Figure 5.23 Equalizer gain when the signal bandwidth exceeds the channel coherence bandwidth [67]. ©2012 IEEE

shows that if there is sufficient margin in the link budget, a considerable amount of signal bandwidth can be accommodated in the narrow channel. Practical bandwidth limitations, such as the small modulation bandwidth of an LED or the narrow coherence bandwidth of the optical wireless channel, can be overcome at the expense of higher computational complexity and an increased SNR requirement. Given that there are sufficient computational resources in the hardware, channel equalization is shown to be key to the realization of OWC links at very high data rates. Fig. 5.23 demonstrates that a 100-times larger throughput at 1 bit/s/Hz average spectral efficiency can be achieved by means of DCO-OFDM with bit and power loading at the expense of an 8.5-dB increase in the electrical SNR requirement. Given the few tens of MHz bandwidths of off-the-shelf LEDs [91, 92] and comparable channel coherence bandwidths [21, 70], an 8.5 dB electrical link margin is shown to enable the transmission of GHz signals with Gbps data throughput on a single link.

5.5.4 Maximum spectral efficiency without an average optical power constraint

The highest spectral efficiency of the transmission schemes for OWC is obtained when the average optical power constraint is relaxed. In order to obtain the electrical SNR requirement when the DC-bias power is not counted towards the signal power in a non-flat dispersive channel, the equalizer gain from Fig. 5.23 needs to be subtracted from the electrical SNR requirement from Fig. 5.4. Since single-carrier transmission incurs a significantly larger penalty in the equalization process, multi-carrier transmission is expected to deliver the highest throughput. When the DC-bias power is counted towards

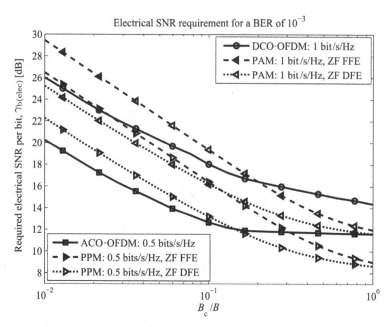

Figure 5.24 Required electrical SNR per bit for a 10^{-3} BER over a 10-dB dynamic range when the signal bandwidth exceeds the channel coherence bandwidth [67]. ©2012 IEEE

the signal power, the DC-bias gain from Fig. 5.22 also needs to be subtracted from the electrical SNR requirement. The result is presented in Fig. 5.24. It is shown that optimum signal clipping allows O-OFDM to close the gap in single-carrier transmission down to 2 dB at $B_c/B = 1$. However, when the signal bandwidth exceeds the channel coherence bandwidth in a dispersive channel, ACO-OFDM shows a lower electrical SNR requirement as compared to PPM with both FFE and DFE. Equivalently, DCO-OFDM is shown to have a lower SNR requirement than PAM with FFE, and it approaches the SNR requirement of PAM with DFE.

By fixing the electrical SNR requirement, the relative performance of the systems can be obtained in terms of spectral efficiency. This is illustrated in Fig. 5.25 for $\gamma_{b(elec)}$ = 25 dB as the signal bandwidth exceeds the channel coherence bandwidth. When the DC power is not counted towards the electrical signal power, DCO-OFDM and ACO-OFDM show superior spectral efficiency in the dispersive optical wireless channel as compared to PAM and PPM, respectively. When the DC power is included in the calculation of the electrical SNR, ACO-OFDM still outperforms PPM. DCO-OFDM outperforms PAM with FFE, and it approaches the performance of PAM with DFE. However, it has to be noted that the analysis of PAM with DFE represents an upper bound for the performance that is achieved when an infinite number of channel taps are considered in the equalizer. In a practical indoor optical wireless channel, where the impulse response only changes slowly, the channel taps and the required bit and power loading parameters with optimum signal biasing can be pre-computed and stored in look-up tables in the memory. Therefore, the computational complexity at the receiver

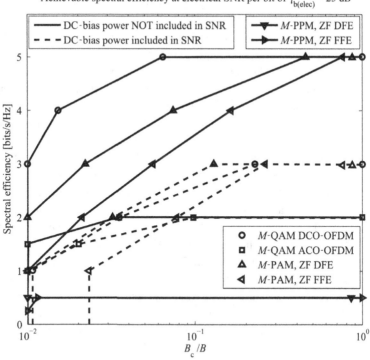

Figure 5.25 Spectral efficiency at an available electrical SNR per bit of 25 dB for a 10^{-3} BER when the signal bandwidth exceeds the channel coherence bandwidth [67]. ©2012 IEEE

comes from the convolution operation of the DFE equalizer in single-carrier transmission and the FFT operation in multi-carrier transmission. It has been shown in [45] that the most efficient DFE implementation requires one FFT and one inverse FFT (IFFT) operation for N channel taps. Therefore, for a fixed FFT size, O-OFDM is expected to require half of the computational complexity of single-carrier transmission with DFE.

5.6 Summary

In this chapter, the multi-carrier modulation schemes DCO-OFDM and ACO-OFDM have been compared with the single-carrier techniques PPM and PAM in terms of the electrical SNR requirement and the spectral efficiency. Aiming at the realization of OWC with very high data rates, the signal bandwidth was increased beyond the coherence bandwidth of the optical wireless channel. Therefore, the modulation schemes for OWC have been compared in a flat fading channel with AWGN, as well as in a dispersive channel. In addition, practical constraints imposed by the optical front-end have been considered, i.e. average electrical power constraint, as well as minimum, average, and maximum optical power constraints.

First, the comparison of the modulation schemes for OWC was presented for the flat fading channel with AWGN. While the single-carrier signals were able to fit within the limited positive linear dynamic range of the transmitter without distortion, the multi-carrier Gaussian signals required optimization of the biasing setup. The optimum biasing setup in the flat fading channel was defined as the DC bias and the signal standard deviation which achieved the minimum electrical SNR requirement for a target BER for a given combination of optical power constraints and a desired QAM order, M. When the average optical power constraint was relaxed, PAM and DCO-OFDM demonstrated the highest spectral efficiency when the DC-bias power was neglected. Including the DC-bias power in the calculation of the electrical SNR, PAM incurred a lower DC-bias penalty, and outperformed DCO-OFDM by 4 dB to 5.5 dB with an increase of the spectral efficiency over a 10-dB dynamic range. Furthermore, the average optical power constraint has been imposed and varied over dynamic ranges of 10 dB, 20 dB, and 30 dB. Neglecting the DC-bias power, PAM and DCO-OFDM maintained their highest spectral efficiency. When DC-bias power was considered, an optical power sweep over more than 50% and 25% of the dynamic range has been shown to be accommodated within an electrical SNR margin of 3 dB for DCO-OFDM and ACO-OFDM, respectively. Because of the Gaussian and half-Gaussian signal distributions, DCO-OFDM demonstrated an absolute minimum SNR requirement towards the middle of the dynamic range, while ACO-OFDM proved to be most suitable for low optical power applications. Nonetheless, DCO-OFDM showed a lower SNR requirement for similar spectral efficiency as compared to ACO-OFDM over more than 85% of average optical power levels with an increase of the dynamic range. However, PAM outperformed DCO-OFDM by 3 dB to 5 dB with an increase of the spectral efficiency over a 20-dB dynamic range. In this case, a transmitter front-end with a wide linear dynamic range of 20 dB or higher has been shown to provide sufficient electrical power to maintain OWC with optical power output close to the boundaries of the dynamic range, where the LED appeared to be off or driven close to its maximum.

In the next section, the maximization of the information rate of the O-OFDM schemes was studied under the constraints of a practical optical front-end with AWGN. Since the non-linear distortion merely resulted in a modification of the electrical SNR, it was directly translated into degradation of the mutual information in the Shannon framework. Here, the minimization of non-linear signal distortion, i.e. maximization of the received electrical SNR and maximization of the information rate, was formulated as an optimization problem. In addition to the minimum and maximum optical power constraints of a linearized dynamic range after pre-distortion, an average electrical power constraint and an average optical power constraint were imposed. The information rate of the two O-OFDM systems was presented, excluding or including the additional DC-bias power in the calculation of the electrical SNR. Considering the DC-bias penalty in the optimization of the biasing setup, an optimum biasing setup has been shown to minimize the SNR penalty for a given average optical power constraint. DCO-OFDM demonstrated a superior information rate to ACO-OFDM over the majority of optical power levels with an increase of the dynamic range or the SNR target. The results could be considered as a lower bound on the capacity of DCO-OFDM and

ACO-OFDM for a given set of optical power constraints and an average electrical power constraint. Neglecting the DC-bias penalty, DCO-OFDM has been shown to achieve Shannon capacity for any average optical power level with minimization of the non-linear distortion, while ACO-OFDM exhibited a 3-dB gap which increases with higher information rate targets. Thus, DCO-OFDM has been proven to deliver the highest throughput in applications where the additional DC-bias power required to create a non-negative signal has been employed to serve a complementary functionality, such as illumination in VLC systems, and it has been excluded from the calculation of the electrical SNR. In IR communication systems, where the DC power has been generally constrained by eye safety regulations, and it has been included in the calculation of the electrical SNR, the optimum signal scaling, and DC-biasing enabled O-OFDM to minimize the SNR penalty.

Finally, O-OFDM with bit and power loading has been compared with PPM and PAM with equalization in terms of the electrical SNR requirement and spectral efficiency in the dispersive optical wireless channel with AWGN. Here, the signal bandwidth has been increased beyond the channel coherence bandwidth. When the additional DC-bias power was neglected, DCO-OFDM with an optimized biasing setup and PAM demonstrated the greatest spectral efficiency in a flat fading channel for the SNR region beyond 6.8 dB. However, since O-OFDM with bit and power loading experienced a lower SNR penalty than PAM with non-linear equalization as the signal bandwidth exceeded the coherence bandwidth of the dispersive optical wireless channel, DCO-OFDM demonstrated the highest spectral efficiency of all the considered modulation schemes for OWC. When the DC-bias power was counted towards the electrical signal power, DCO-OFDM and ACO-OFDM experienced a greater SNR penalty due to the DC bias as compared to PAM and PPM, respectively. However, the presented optimum signal shaping framework enabled O-OFDM to greatly reduce this penalty and to minimize the gap to single-carrier transmission within 2 dB for comparable signal and channel bandwidths. Here, DCO-OFDM and ACO-OFDM with optimum biasing showed an improvement of 6.5 dB and 1.4 dB, respectively, as compared to a conventional minimum-distortion front-end biasing setup with signal clipping at 4σ. When the signal bandwidth exceeded the channel coherence bandwidth, DCO-OFDM outperformed PAM with linear equalization, and it approached the spectral efficiency of the more computationally intensive PAM with non-linear equalization, while ACO-OFDM outperformed PPM with linear as well as non-linear equalization.

6 MIMO transmission

6.1 Introduction

In order to provide sufficient illumination, light installations are typically equipped with multiple light emitting diodes (LEDs). This property can readily be exploited to create optical multiple-input–multiple-output (MIMO) communication systems. MIMO techniques are well-established and widely implemented in many radio frequency (RF) systems as they offer high data rates by increasing spectral efficiency [184, 185]. Off-the-shelf LEDs provide only a limited bandwidth of about 30–50 MHz for incoherent infrared (IR) light and even less for visible light. Consequently, these incoherent light sources restrict the available bandwidth of practical optical wireless communication (OWC) systems. Therefore, it is equally important to achieve high spectral efficiencies in OWC [186]. For free-space optical (FSO) transmission, the effects of MIMO have already been studied. It has been shown that spatial diversity can combat the fading effects due to scattering and scintillation caused by atmospheric turbulence [187, 188]. Ongoing research activities intend to increase the capacity of indoor OWC systems by MIMO techniques [189, 190]. However, for indoor OWC it is still not clear to what extent MIMO techniques can provide gains. This is because in indoor environments there are no fading effects caused by turbulence, especially if line-of-sight (LOS) scenarios are considered. Therefore, indoor optical wireless links are envisaged to be highly correlated, enabling only minor diversity gains. As MIMO techniques mostly rely on spatially uncorrelated channels, it is unclear whether the optical propagation channel in indoor environments can offer sufficiently low channel correlation.

By employing multiple LEDs as transmitters (Txs) and multiple photodiodes (PDs) as receivers (Rxs) the OWC system benefits from multiplexing and diversity gains to increase the throughput. In this chapter, the performances of several MIMO techniques for OWC are demonstrated assuming LOS channel conditions [191]. These conditions are considered as a worse case scenario because of the high channel correlation due to spatial symmetries and very little path gain differences. Specifically, several 4×4 setups with different transmitter spacings and different positions of the receiver array are considered. The following MIMO algorithms are compared: repetition coding (RC) [188], spatial multiplexing (SMP), and spatial modulation (SM) [192–195]. Particularly, the bit-error ratio (BER) performance of these MIMO techniques is approximated analytically, and the theoretical BER bounds are verified by means of Monte Carlo simulations. The results show that due to diversity gains,

RC is robust to various transmitter–receiver alignments. However, as RC does not provide spatial multiplexing gains, it requires large signal constellation sizes to enable high spectral efficiencies. In contrast, SMP enables high data rates by exploiting multiplexing gains. In order to provide these gains, SMP needs sufficiently low channel correlation. SM is a combined MIMO and digital modulation technique which is shown to be more robust to high channel correlation compared to SMP, while enabling larger spectral efficiency compared to RC. Furthermore, the computational complexity of the MIMO techniques is compared as a function of the number of transmitters and receivers and the modulation order [196]. Even though SM has lower achievable spectral efficiency than SMP, SM is shown to have considerably lower computational complexity. In addition, the effect of induced power imbalance between the multiple transmitters is studied. It is found that power imbalance can substantially improve the performance of both SMP and SM as it reduces channel correlation. It is shown that blocking some of the links is an acceptable method to reduce channel correlation. Even though the blocking reduces the received energy, it outweighs the signal-to-noise ratio (SNR) degradation by providing improved channel conditions for SMP and SM. For example, blocking 4 of the 16 links of the 4×4 setup improves the BER performance of SMP by more than 20 dB, while the effective SNR is reduced by about 2 dB due to the blocking.

6.2   System model

An optical wireless MIMO transmission system is realized through intensity modulation and direct detection (IM/DD) of an optical carrier by means of multiple incoherent light sources and multiple photodetectors. In particular, the system is equipped with N_t LEDs at the transmitter side and N_r PDs at the receiver side. The conventional model for a MIMO transmission link over a single-tap channel is employed as follows:

$$\mathbf{y} = \mathbf{H}\,\mathbf{x} + \mathbf{w}, \tag{6.1}$$

where the received signal vector, $\mathbf{y}$, the channel matrix, $\mathbf{H}$, the transmitted signal vector, $\mathbf{x}$, and the additive white Gaussian noise (AWGN) vector, $\mathbf{w}$, are real-valued. In this study, unipolar multi-level pulse amplitude modulation (M-PAM) is employed, where M denotes the signal constellation size. On a single link, M-PAM achieves a spectral efficiency of $\log_2(M)$ bits/s/Hz. PAM is selected because it is more bandwidth efficient compared to other pulse modulation techniques such as on–off keying (OOK), pulse-width modulation (PWM), and pulse-position modulation (PPM). Moreover, PAM has been shown to have a slightly better electrical power efficiency in the flat fading channel without dispersion (which is assumed in this chapter) when compared to direct-current-biased optical orthogonal frequency division multiplexing (DCO-OFDM) as shown in Chapter 5. As a result, the AWGN has an electrical power spectral density (PSD) of $N_0/2$ and an electrical power of $\sigma_{\mathrm{AWGN}}^2 = BN_0/2$ over a bandwidth of B. Based on the front-end setup, the electrical power of the AWGN can be obtained from (2.1). Non-linear distortion is not assumed at the transmitter side for reasons of simplicity.

The transmitted signal vector is expressed as $\mathbf{x} = \begin{bmatrix} x_1 & \ldots & x_{N_t} \end{bmatrix}^T$, with $[\cdot]^T$ being the transpose operator. The elements of $\mathbf{x}$ indicate which signal is emitted by each optical transmitter, i.e. x_{n_t} denotes the signal emitted by transmitter n_t. The $N_r \times N_t$ channel matrix $\mathbf{H}$ is expressed as follows:

$$\mathbf{H} = \begin{pmatrix} h_{11} & \cdots & h_{1N_t} \\ \vdots & \ddots & \vdots \\ h_{N_r 1} & \cdots & h_{N_r N_t} \end{pmatrix}, \qquad (6.2)$$

where $h_{n_r n_t}$ represents the optical channel transfer factor, i.e. the optical path gain, of the wireless link between transmitter n_t and receiver n_r. As the LEDs are in close proximity, they can be jointly driven by exactly the same baseband hardware and electronic driver. Therefore, the transmission is assumed to be perfectly synchronized. In addition, there is only a very small path difference between the multiple transmitter–receiver links of some centimeter as shown in Section 6.4. Therefore, there is negligible temporal delay between the multiple links, and consequently the system model given in (6.1) is considered as a single-tap channel without time dispersion.

In this study, optical wireless links with LOS channel characteristics are assumed. Figure 6.1 (right-hand side) illustrates a directed LOS link. As shown, $\theta_{\text{Tx,d}}$ is the observation angle of the transmitter on the direct path, and $\theta_{\text{Rx,d}}$ is the incident angle of the receiver on the direct path. Furthermore, d represents the distance between the transmitter and the receiver. Using this geometric scenario, the path gain of an optical propagation link can be calculated as follows:

$$h = \frac{n_{\text{Tx}} + 1}{2\pi} \cos^{n_{\text{Tx}}}(\theta_{\text{Tx,d}}) \frac{A}{d^2} \cos(\theta_{\text{Rx,d}}) \text{rect}(\theta_{\text{Rx,d}}) S_{\text{PD}} G_{\text{TIA}} T_{\text{OF}} G_{\text{OC}} / \sqrt{R_{\text{load}}} . \qquad (6.3)$$

The field of view (FOV) semi-angle of a single LED, $\theta_{\text{FOV,Tx}}$, and the FOV semi-angle of a single PD, $\theta_{\text{FOV,Rx}}$, are assumed to be $15°$. By the use of (2.26) and $\theta_{\text{FOV,Tx}}$, the Lambertian mode number of the LED, n_{Tx}, can be obtained. These FOVs have been chosen with regard to a practical LOS indoor OWC system which has been developed and implemented within the European project OMEGA [197, 198]. The area of a single PD, A, is assumed to be 1 cm^2. The additional parameters of the receiver front-end, i.e. the responsivity of a PD, S_{PD}, the gain of a transimpedance amplifier (TIA), G_{TIA}, the

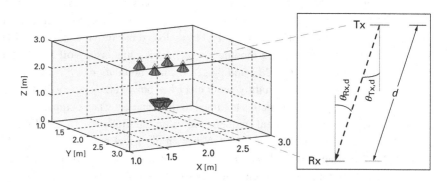

Figure 6.1 Geometric scenario used for calculation of channel coefficients [191]. ©2013 IEEE

transmittance of an optical filter, T_{OF}, the gain of an optical concentrator, G_{OC}, and the load resistance, R_{load}, are assumed to be unity. Clearly, the channel coefficient $h_{n_r n_t}$ depends on the specific position of LED n_t and PD n_r. If PD n_r is not within the FOV of LED n_t and vice versa, then the respective path gain coefficient $h_{n_r n_t}$ is zero.

In the following, a 4×4 indoor scenario ($N_r = 4$ and $N_t = 4$) is considered, and it is located within a 4.0 m $\times$ 4.0 m $\times$ 3.0 m (X, Y, Z) room. The transmitters are placed at a height of $z = 2.50$ m, and they are oriented to point straight down from the ceiling. The receivers are located at a height of $z = 0.75$ m (e.g. height of a table) and are oriented to point straight up to the ceiling. Both transmitters and receivers are aligned in a quadratical 2×2 array, which is centered in the middle of the room. On the basis of this scenario, different static setups are investigated with different spacings of the individual transmitters on the X- and Y-axis, denoted by d_{Tx}. The spacing of the receivers is assumed to be 0.1 m on the X- and Y-axis for all considered setups. This receiver spacing would enable the implementation of the receiver array into typical laptop computers. Figure 6.1 (left-hand side) shows the positioning of the 4×4 setup. The receivers are displayed as dots and the transmitters as triangles. The plotted cones represent the orientation of the transmit beams and the orientation of the FOVs of the receivers, respectively. The cone angles are related to the FOV semi-angles of the transmitter and receiver devices.

6.3   MIMO techniques

There are several MIMO techniques which are suitable candidates for indoor OWC. In this chapter, the following MIMO schemes are studied and compared in terms of computational complexity and BER performance: RC, SMP, and SM. For this purpose, first, the detection algorithm needs to be defined. All considered MIMO techniques are assumed to perform maximum likelihood sequence detection (MLSD) at the receiver side with perfect knowledge of the channel and ideal time synchronization. The decoder decides for the signal vector $\widehat{\mathbf{x}}$, which minimizes the Euclidean distance between the actual received signal vector $\mathbf{y}$ and all potential received signals as follows:

$$\widehat{\mathbf{x}} = \arg\max_{\mathbf{x}} p_{\mathbf{y}}(\mathbf{y}|\mathbf{x}, \mathbf{H}) = \arg\min_{\mathbf{x}} \|\mathbf{y} - \mathbf{H}\mathbf{x}\|_F^2, \tag{6.4}$$

where $p_{\mathbf{y}}$ is the probability density function of $\mathbf{y}$ conditioned on $\mathbf{x}$ and $\mathbf{H}$. Here, $\|\cdot\|_F$ denotes the Frobenius norm. In the following, the signaling in RC, SMP, and SM is discussed in detail, and the respective analytical BER bounds are presented.

6.3.1   Repetition coding

The simplest MIMO transmission technique is RC which simultaneously emits the same signal from all transmitters. Therefore, in RC $x_1 = x_2 = \cdots = x_{N_t}$. RC is known to achieve good performance in free-space OWC because of transmit diversity [188]. In [199], it is shown that RC can outperform orthogonal space-time block codes (OSTBC) such as the Alamouti scheme [200] and single-input–multiple-output (SIMO) setups. This is due to the fact that the intensities coming from the several transmitters constructively add up at the receiver side. When it comes to evaluation of the performance of multiple-LED transmission schemes for OWC, it is important that the average optical power of the transmitter is constrained equally among the considered schemes. Here, an infinite non-negative linear dynamic range of the transmitter is considered under an average optical power constraint. In this case, a common metric used to evaluate the performance of OWC systems is the optical SNR at the transmitter side defined as $P_{s(opt)}/\sigma_{AWGN}$ [171, 201]. The squared metric, $P_{s(opt)}^2/\sigma_{AWGN}^2$, is further adopted and often referred to as the electrical SNR at the transmitter side [194]. The latter approach is considered in this study. Without loss of generality, rectangular pulses are employed in conjunction with M-PAM, which is the considered modulation format for comparison of the MIMO schemes. The optical power levels of the PAM symbols are given as follows:

$$P_{opt,p}^{PAM} = \frac{2\,P_{s(opt)}}{M-1}\,p\,, \quad \text{for } p = 0, 1, \ldots, (M-1), \tag{6.5}$$

where $P_{s(opt)}$ is the average optical symbol power. The expression from (4.12) can be used as a lower bound of the BER performance of M-PAM on a single link as follows:

$$
\begin{aligned}
\text{BER}_{PAM} &\geq \frac{2\,(M-1)}{M\,\log_2(M)}\,Q\left(\sqrt{\frac{6}{M^2-1}\frac{h^2 P_{s(opt)}^2 T_s}{N_0}F_{OE}G_{DC}}\right) \\
&\geq \frac{2\,(M-1)}{M\,\log_2(M)}\,Q\left(\frac{1}{M-1}\sqrt{h^2 \text{SNR}_{Tx}}\right),
\end{aligned}
\tag{6.6}
$$

where $\text{SNR}_{Tx} = 2P_{s(opt)}^2 T_s/N_0$ is the electrical SNR at the transmitter side considered in this study, and $Q(\cdot)$ is the complementary cumulative distribution function (CCDF) of a standard normal distribution. The symbol duration is denoted by T_s. The optical-to-electrical (O/E) conversion factor in M-PAM, can be obtained from (4.3) in Section 4.3.2 as follows: $F_{OE} = P_{s(elec)}/P_{s(opt)}^2$. In addition, the DC-bias penalty, G_{DC}, is given for M-PAM in (4.10).

As RC simultaneously emits the same signal from several LEDs, the optical transmission power is equally distributed across all emitters. Thus, the optical intensities given in (6.5) have to be divided by factor N_t. By doing so, the mean optical power emitted is constant, irrespective of the number of employed transmitters. This ensures comparability of different setups and transmission techniques. Thus, the BER of RC for an arbitrary $N_r \times N_t$ setup can be generalized from the BER of M-PAM given in (6.6) as follows:

$$\text{BER}_{RC} \geq \frac{2\,(M-1)}{M\,\log_2(M)}\,Q\left(\frac{1}{M-1}\sqrt{\frac{\text{SNR}_{Tx}}{N_t^2}\sum_{n_r=1}^{N_r}\left(\sum_{n_t=1}^{N_t}h_{n_r n_t}\right)^2}\right). \tag{6.7}$$

The intensities emitted by the multiple transmitters constructively add up at the receiver side leading to an effective received optical symbol power of $P_{\tilde{s}(\text{opt})} = \sum_{n_t=1}^{N_t} \frac{P_{s(\text{opt})}}{N_t} h_{n_r n_t}$ at receiver n_r, including the parameters of its front-end through the channel gains. Consequently, the individual channel gains $h_{n_r n_t} \in [0; 1]$ induce a distinctive attenuation of the transmitted signals (path loss), depending on the specific link characteristics. The N_r received signals are combined by maximum ratio combining (MRC) [202]. Thus, by applying MRC, the received signals with a high SNR are weighted more than signals with a low SNR. Consequently, the electrical SNR at the receiver side after the combiner becomes:

$$
\begin{aligned}
\text{SNR}_{\text{Rx}} &= \frac{2T_s}{N_0} \sum_{n_r=1}^{N_r} P_{\tilde{s}(\text{opt})}^2 \\
&= \frac{2T_s}{N_0} \sum_{n_r=1}^{N_r} \left(\sum_{n_t=1}^{N_t} \frac{P_{s(\text{opt})}}{N_t} h_{n_r n_t} \right)^2 \\
&= \frac{\text{SNR}_{\text{Tx}}}{N_t^2} \sum_{n_r=1}^{N_r} \left(\sum_{n_t=1}^{N_t} h_{n_r n_t} \right)^2,
\end{aligned}
\tag{6.8}
$$

which corresponds to the SNR given in the argument of the Q function in (6.7). Consequently, the BER of RC is only affected by the transfer factors of the optical wireless links and the received optical power. Thus, RC can be represented by a simple single-input–single-output (SISO) scheme which provides the same received electrical energy.

6.3.2 Spatial multiplexing

Another well-known MIMO technique is SMP. In SMP, independent data streams are simultaneously emitted from all transmitters. Therefore, SMP provides an enhanced spectral efficiency of $N_t \log_2(M)$ bits/s/Hz. Similar to RC, SMP employs M-PAM. The optical power is equally distributed across all emitters to ensure that both schemes use the same mean transmission power. In SMP, the signal vector $\mathbf{x}$ has N_t elements which are independent M-PAM modulated signals according to (6.5), where the optical power levels, $P_{\text{opt},p}^{\text{PAM}}$, are divided by N_t. Provided that the SMP receiver performs MLSD, the pairwise error probability (PEP) is the probability that the receiver mistakes the transmitted signal vector $\mathbf{x}_{p^{(1)}}$ for another vector $\mathbf{x}_{p^{(2)}}$, given knowledge of the channel matrix $\mathbf{H}$. Therefore, the PEP of SMP can be expressed as follows:

$$
\begin{aligned}
\text{PEP}_{\text{SMP}} &= \text{PEP}(\mathbf{x}_{p^{(1)}} \rightarrow \mathbf{x}_{p^{(2)}} | \mathbf{H}) \\
&= Q\left(\sqrt{\frac{T_s}{2N_0}} \left\| \mathbf{H}\left(\mathbf{x}_{p^{(1)}} - \mathbf{x}_{p^{(2)}} \right) \right\|_F^2 \right).
\end{aligned}
\tag{6.9}
$$

Using the PEP and considering all M^{N_t} possible combinations of the transmitted signal vector, the BER of SMP can be approximated by union bound methods. The upper bound is given as follows:

$$
\mathrm{BER_{SMP}} \leq \frac{1}{M^{N_t} \log_2(M^{N_t})} \sum_{p^{(1)}=1}^{M^{N_t}} \sum_{p^{(2)}=1}^{M^{N_t}} d_\mathrm{H}(b_{p^{(1)}}, b_{p^{(2)}})
$$

$$
\times Q\left(\sqrt{\frac{T_s}{2N_0}} \left\| \mathbf{H}\left(\mathbf{x}_{p^{(1)}} - \mathbf{x}_{p^{(2)}}\right) \right\|_\mathrm{F}^2 \right),
$$

(6.10)

where $d_\mathrm{H}(b_{p^{(1)}}, b_{p^{(2)}})$ denotes the Hamming distance of the two bit assignments $b_{p^{(1)}}$ and $b_{p^{(2)}}$ of the signal vectors $\mathbf{x}_{p^{(1)}}$ and $\mathbf{x}_{p^{(2)}}$. For instance, if a setup with $N_t = 4$ and $M = 2$ is assumed, the bit sequence "1 0 0 1" is assigned to $\mathbf{x}_{10} = [P_{s(\mathrm{opt})}/2\ 0\ 0\ P_{s(\mathrm{opt})}/2]^T$ and "1 0 0 0" is assigned to $\mathbf{x}_9 = [P_{s(\mathrm{opt})}/2\ 0\ 0\ 0]^T$, resulting in $d_\mathrm{H}(b_{10}, b_9) = 1$. Therefore, $d_\mathrm{H}(\cdot, \cdot)$ states the number of bit errors when erroneously detecting $\mathbf{x}_{p^{(2)}}$ at the receiver side instead of the actually transmitted signal vector $\mathbf{x}_{p^{(1)}}$.

6.3.3   Spatial modulation

Finally, SM is also considered. SM is a combined MIMO and digital modulation technique, proposed in [192] and further investigated in [193, 194, 203, 204]. In SM, the conventional signal constellation diagram is extended to an additional dimension, namely the spatial dimension, which is used to transmit additional bits. A unique binary sequence, i.e. the spatial symbol, is assigned to each transmitter in the transmitting array. A transmitter is only activated when the random spatial symbol to be transmitted matches the pre-allocated spatial symbol of the transmitter. Thus, only one transmitter is activated at any PAM symbol duration. Therefore, there is only one non-zero element in the signal vector $\mathbf{x}$ to be transmitted at a time instant, and this element is the digitally modulated signal. The index of the non-zero element is the spatial symbol. Since SM simultaneously transmits data in the signal domain and the spatial domain, it provides an enhanced spectral efficiency of $\log_2(N_t) + \log_2(M)$ bits/s/Hz. Moreover, as only one transmitter is activated at any symbol duration, SM completely avoids inter-symbol interference (ISI) across the different channels. Thus, SM has lower decoding complexity compared to other MIMO schemes [205–207]. Due to the distinct channel transfer factors between a particular transmitter and the receiver, the receiver is able to detect which transmitter is activated and hence is able to detect the spatial symbol. Fig. 6.2 illustrates the operation of SM in a setup with $N_t = 4$ LEDs and a signal constellation size of $M = 4$. The bits to be transmitted are passed to the SM encoder, which maps them to the respective signal and transmitter index. In this example, the last two bits denote the index of the transmitter that emits the signal, while the first two bits represent the actual signal to be transmitted. For instance, the bit sequence "1 1 1 0" is represented by transmitter number 4 emitting signal $P_{\mathrm{opt},3}$. In contrast to RC and SMP, signals with intensity $P_{\mathrm{opt}} = 0$ cannot be used for the signal modulation in SM. This is because in this case no transmitter will be active and the spatial information will be lost. Therefore, the intensities of M-PAM given in (6.5) have to be modified to be suitable

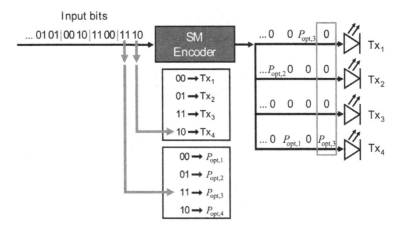

Figure 6.2 Illustration of SM operation with $N_t = 4$ and $M = 4$ [191]. The first two bits in the block of four bits determine the PAM symbol and the second two bits determine the active LED. ©2013 IEEE

for SM as follows:

$$P_{\text{opt},p}^{\text{SM}} = \frac{2\,P_{s(\text{opt})}}{M+1}\,p \quad \text{for } p = 1\ldots M. \tag{6.11}$$

Consequently, the minimum distance between two SM signals is $2P_{s(\text{opt})}/(M+1)$, while the minimum distance in M-PAM is $2\,P_{s(\text{opt})}/(M-1)$. The smaller signal distance of SM might induce poorer BER performance because the error probability depends on the Euclidean distance of the transmitted signals. However, as SM additionally encodes data bits in the spatial domain, it can provide the same spectral efficiency as M-PAM with a lower signal constellation size, hence effectively enlarging the distance of the signal points. As the SM receiver has to detect the transmitter that has sent the signal, its performance depends on the differentiability of the multiple channels. Therefore, the performance of SM is affected by channel correlation. The PEP of SM is given as follows:

$$\text{PEP}_{\text{SM}} = \text{PEP}(\mathbf{x}_{p^{(1)}} \rightarrow \mathbf{x}_{p^{(2)}} | \mathbf{H})$$

$$= Q\left(\sqrt{\frac{T_s}{2\,N_0}}\,\left\|\mathbf{H}\left(\mathbf{x}_{p^{(1)}} - \mathbf{x}_{p^{(2)}}\right)\right\|_F^2\right) \tag{6.12}$$

$$= Q\left(\sqrt{\frac{T_s}{2\,N_0}\sum_{n_r=1}^{N_r}\left|P_{\text{opt},p^{(2)}}^{\text{SM}}\,h_{n_r n_t^{(2)}} - P_{\text{opt},p^{(1)}}^{\text{SM}}\,h_{n_r n_t^{(1)}}\right|^2}\right).$$

This denotes the probability that the receiver decides for intensity $P_{\text{opt},p^{(2)}}^{\text{SM}}$ to be emitted by transmitter $n_t^{(2)}$, when actually transmitter $n_t^{(1)}$ has emitted intensity $P_{\text{opt},p^{(1)}}^{\text{SM}}$. Using this PEP and considering all possible MN_t signal combinations, the BER of SM can be approximated by union bound methods. The upper bound is given as follows:

$$
\text{BER}_{\text{SM}} \le \frac{1}{MN_t \log_2(MN_t)} \sum_{p^{(1)}=1}^{M} \sum_{n_t^{(1)}=1}^{N_t} \sum_{p^{(2)}=1}^{M} \sum_{n_t^{(2)}=1}^{N_t} d_{\text{H}} \left(b_{p^{(1)} n_t^{(1)}}, b_{p^{(2)} n_t^{(2)}} \right)
$$

$$
\times Q \left(\sqrt{ \frac{T_s}{2 N_0} \sum_{n_r=1}^{N_r} \left| P_{\text{opt},p^{(2)}}^{\text{SM}} \, h_{n_r n_t^{(2)}} - P_{\text{opt},p^{(1)}}^{\text{SM}} \, h_{n_r n_t^{(1)}} \right|^2 } \right),
$$

(6.13)

where $b_{p^{(1)} n_t^{(1)}}$ is the bit assignment which is conveyed, when intensity $P_{\text{opt},p^{(1)}}^{\text{SM}}$ is emitted by transmitter $n_t^{(1)}$, and $b_{p^{(2)} n_t^{(2)}}$ is the bit assignment which is encoded when intensity $P_{\text{opt},p^{(2)}}^{\text{SM}}$ is emitted by transmitter $n_t^{(2)}$. Consequently, $d_{\text{H}}(b_{p^{(1)} n_t^{(1)}}, b_{p^{(2)} n_t^{(2)}})$ states the number of bit errors when erroneously decoding the bit sequence $b_{p^{(2)} n_t^{(2)}}$ at the receiver side instead of the actually transmitted sequence $b_{p^{(1)} n_t^{(1)}}$.

6.3.4   Computational complexity

On the basis of the MLSD algorithm, the computational complexity of optical SM is analyzed and compared with the complexity of RC and SMP. Both RC and SMP are common optical MIMO techniques. RC is the simplest MIMO transmission technique as it simultaneously emits the same signal from all optical transmitters. By applying SMP, independent data streams are simultaneously emitted from all transmitters. The computational complexity is defined as the total number of required mathematical operations, i.e. multiplications, additions, and subtractions that are required for MLSD. The ascertained numbers are listed in Table 6.1. It can be seen that RC and SMP need the same number of operations to provide equal spectral efficiency, SE, whereas the detection of SM is less computationally intensive. This is due to the fact that SM avoids ISI across the different channels, as only one transmitter is active at any symbol duration in contrast to SMP. Moreover, as SM conveys additional bits in the spatial domain, it

Table 6.1 Comparison of spectral efficiencies and the computational complexity at the receiver for different MIMO techniques

MIMO technique	Spectral efficiency SE in bits/s/Hz	Number of mathematical operations at receiver
RC	$\log_2(M)$	$2^{\text{SE}}(2 N_t N_r + N_r - 1)$ $= M(2 N_t N_r + N_r - 1)$
SMP	$N_t \log_2(M)$	$2^{\text{SE}}(2 N_t N_r + N_r - 1)$ $= M^{N_t}(2 N_t N_r + N_r - 1)$
SM	$\log_2(MN_t)$	$2^{\text{SE}}(3 N_r - 1)$ $= MN_t(3 N_r - 1)$

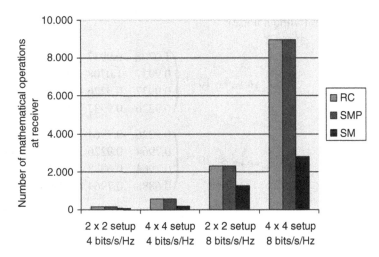

Figure 6.3 Computational complexity at receiver of RC, SMP, and SM for different setups and spectral efficiencies [196]. ©2013 IEEE

can achieve the same spectral efficiency as RC, but with a reduced signal constellation size. Thus, for SM, there are less signal constellation points to be considered when performing the MLSD. For instance, if a setup with $N_t = 4$ and $N_r = 4$ is assumed, which provides a spectral efficiency of SE = 4 bits/s/Hz, RC and SMP both require 560 operations, whereas SM requires only 176 operations. Fig. 6.3 displays the computational complexity of RC, SMP, and SM for different setups and spectral efficiencies. Consequently, as SM requires much lower computational complexity compared to common MIMO techniques, it enables the implementation of efficient and simple optical wireless MIMO receivers.

6.4 BER performance

In this section, the BER performance of the considered MIMO techniques is assessed. Several setup scenarios are studied using the system model from Section 6.2. These include varying the separation of transmitters, varying the position of receivers, power imbalance between transmitters, and link blockage. In order to ensure comparability, the mean emitted optical power is the same for all the MIMO techniques in each scenario. The specific path loss caused by the particular distance and angular alignment of the individual transmitters and receivers is taken into account for each setup. Therefore, the BER at the receiver side is evaluated with respect to the transmitted SNR, SNR_{Tx}. This is because the BER evaluated through the received SNR would disregard the individual path losses of the different setups, thus disallowing a fair performance comparison.

6.4.1 Varying the separation of transmitters

The 4×4 setup is considered with transmitter spacings on the X- and Y-axis of $d_{Tx} = \{0.2, 0.4, 0.6\}$ m. Applying (6.3) to these scenarios gives the following channel matrices (without noise):

$$\mathbf{H}_{d_{\mathrm{Tx}}=0.2\mathrm{m}} \approx 10^{-4} \begin{pmatrix} 1.0708 & 0.9937 & 0.9937 & 0.9226 \\ 0.9937 & 1.0708 & 0.9226 & 0.9937 \\ 0.9937 & 0.9226 & 1.0708 & 0.9937 \\ 0.9226 & 0.9937 & 0.9937 & 1.0708 \end{pmatrix},$$

$$\mathbf{H}_{d_{\mathrm{Tx}}=0.4\mathrm{m}} \approx 10^{-4} \begin{pmatrix} 0.9226 & 0.7964 & 0.7964 & 0.6888 \\ 0.7964 & 0.9226 & 0.6888 & 0.7964 \\ 0.7964 & 0.6888 & 0.9226 & 0.7964 \\ 0.6888 & 0.7964 & 0.7964 & 0.9226 \end{pmatrix}, \qquad (6.14)$$

$$\mathbf{H}_{d_{\mathrm{Tx}}=0.6\mathrm{m}} \approx 10^{-4} \begin{pmatrix} 0.6888 & 0.5559 & 0.5559 & 0 \\ 0.5559 & 0.6888 & 0 & 0.5559 \\ 0.5559 & 0 & 0.6888 & 0.5559 \\ 0 & 0.5559 & 0.5559 & 0.6888 \end{pmatrix}.$$

The symmetric arrangement of the transmitters and receivers leads to equal channel gains for the links with the same geometric alignment. Moreover, if the spacing between the transmitters is small, the gains are quite similar, while for larger d_{Tx}, the difference between the links increases. When $d_{\mathrm{Tx}} = 0.6$ m, some transmitters fall outside of a receiver's FOV, or vice versa, resulting in $h_{n_r n_t} = 0$. The optical channel coefficients are in the region of 10^{-4} and the electrical path loss at the receiver side is about 80 dB. Since the transmitted SNR is considered, the BER curves displayed in the following figures have an SNR offset of about 80 dB with respect to the received SNR. Furthermore, if the diffuse transmission portion induced by the first-order reflections on the surfaces (walls) is considered, the reflected optical power impinging on the receivers is in the range of $10^{-10} P_{\mathrm{opt}}$ (assuming ideal conditions such as Lambertian reflectors and a reflectivity of $\rho = 1$). This results in an electrical path loss which is about $110 - 120$ dB larger than the path loss of the LOS component. Since the path loss of higher-order reflections is even larger, the reflections can be neglected in the following. Therefore, only the LOS gains given in (6.14) are considered in this study without any diffuse components. However, additional multipath reflections would enhance the differentiability of the MIMO channels, reducing channel correlation. Hence, as there are no reflections, the considered LOS scenario is subject to highly correlated links, and it constitutes the worst case scenario with respect to channel correlation.

In general, a larger transmitter separation, d_{Tx}, increases the path loss. The penalty on the received electrical SNR can be expressed for different transmitter spacings as follows:

$$\Delta_{d_{\mathrm{Tx}}=x}^{\mathrm{SNR_{Rx}}} = 10 \log_{10} \left(\frac{\mathrm{SNR}_{\mathrm{Rx},d_{\mathrm{Tx}}=x}}{\mathrm{SNR}_{\mathrm{Rx},d_{\mathrm{Tx}}=0.2\mathrm{m}}} \right). \qquad (6.15)$$

The received electrical SNR at $d_{\mathrm{Tx}} = x$ m is denoted as $\mathrm{SNR}_{\mathrm{Rx},d_{\mathrm{Tx}}=x}$. The penalty is relative to the $d_{\mathrm{Tx}} = 0.2$ m setup which provides the lowest path loss. The different considered scenarios have the following penalties:

$$\Delta_{d_{Tx}=0.2m}^{SNR_{Rx}} = \quad 0 \text{ dB},$$

$$\Delta_{d_{Tx}=0.4m}^{SNR_{Rx}} \approx -1.88 \text{ dB}, \qquad (6.16)$$

$$\Delta_{d_{Tx}=0.6m}^{SNR_{Rx}} \approx -6.89 \text{ dB}.$$

Besides, the maximum difference in path length of the multiple transmitter–receiver links is about 33.30 mm. This difference results in a maximum delay variation of 111.06 ps. A delay variation of several ps only has an effect when switching speeds in the region of several GHz. Since off-the-shelf LEDs are considered that provide a bandwidth of about 30–50 MHz, this delay variation can be neglected and ideal synchronization of all links is assumed without time dispersion.

Fig. 6.4(a) shows the BER performance of RC, SMP, and SM for the three setup scenarios for a spectral efficiency of SE = 4 bits/s/Hz. For $d_{Tx} = 0.2$ m and $d_{Tx} = 0.4$ m, RC gives the best performance, while SMP performs the worst with a low slope of the BER curve within the depicted SNR_{Tx} range. This is due to the fact, that for both scenarios the channel gains are quite similar providing high channel correlation. Although the performance of SM also depends on the differences between the links, SM is more robust to these channel conditions. Thus, SM provides a lower BER and a steeper slope of the BER curve as compared to SMP. If $d_{Tx} = 0.6$ m, SMP and SM outperform RC at a BER of 10^{-5} by about 10 dB and 9 dB, respectively. RC performs about 7 dB worse as compared to the $d_{Tx} = 0.2$ m case because of the received electrical SNR penalty of $\Delta_{d_{Tx}=0.6m}^{SNR_{Rx}} \approx -6.89$ dB. Despite the larger path loss of the $d_{Tx} = 0.6$ m setup as compared to $d_{Tx} = 0.2$ m and $d_{Tx} = 0.4$ m, SMP and SM with $d_{Tx} = 0.6$ m even outperform RC with $d_{Tx} = 0.2$ m and $d_{Tx} = 0.4$ m. SMP outperforms SM by about 1 dB at high SNR_{Tx}. Because of its multiplexing gain, SMP can operate with a reduced signal constellation size of $M = 2$ as opposed to SM which has to operate with $M = 4$ to provide the same data rate. However, in the low-SNR_{Tx} regions of up to about 103 dB, SM provides the best BER performance. This implies that SM benefits from conveying information in the spatial domain, especially at low SNR_{Tx}. In the high-SNR_{Tx} regions, SM has the disadvantage that it has to use a larger signal constellation size to provide the same spectral efficiency as SMP. Accordingly, SMP requires a high SNR_{Tx} to separate the individual signal streams at the receiver side and to benefit from its multiplexing gain. Moreover, the theoretical lower and upper error bounds (shown by markers) given in (6.7), (6.10), and (6.13) closely match the simulation results (shown by lines). Thus, the error bounds provide an accurate approximation of the BER at high SNR_{Tx}.

Fig. 6.4(b) shows the BER performance of the three schemes in the same setup scenarios, but for an enhanced spectral efficiency of SE = 8 bits/s/Hz. RC requires an SNR_{Tx} increase of about 24 dB to achieve the same BER of 10^{-5} as compared to the case when SE = 4 bits/s/Hz. However, SM outperforms RC up to an SNR_{Tx} of about 130 dB for $d_{Tx} = 0.2$ m and up to an SNR_{Tx} of about 134 dB for $d_{Tx} = 0.4$ m. In the first two scenarios for the transmitter separation, SMP is outperformed by RC and SM. However, if $d_{Tx} = 0.6$ m, SMP shows the best performance as it outperforms SM and

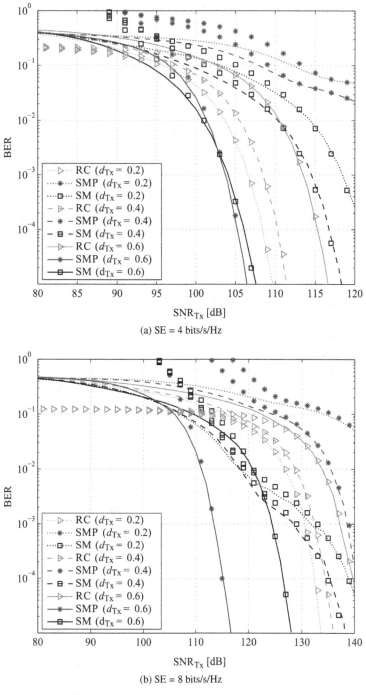

Figure 6.4 Comparison of RC, SMP, and SM with varying distance d_{Tx} of the transmitters on the X- and Y-axis [191]. Spectral efficiencies of SE = 4 bits/s/Hz and SE = 8 bits/s/Hz are considered in a 4×4 setup (lines show simulation results and markers are analytical error bounds). ©2013 IEEE

RC by about 12 dB and 25 dB, respectively. Hence, due to its larger multiplexing gain, SMP requires only 10 dB of additional SNR_{Tx} to provide the same BER performance when doubling the spectral efficiency from 4 to 8 bits/s/Hz, while SM requires 21 dB of additional SNR_{Tx}. As a general observation, RC performs worse when the transmitter spacing is increased, while the opposite trend is evident for SM and SMP due to their multiplexing gain at higher transmitter spacings.

If the transmitter spacing is further increased to $d_{\text{Tx}} = 0.7$ m, the following channel matrix is obtained:

$$\mathbf{H}_{d_{\text{Tx}}=0.7\text{m}} \approx 10^{-4} \begin{pmatrix} 0.5658 & 0 & 0 & 0 \\ 0 & 0.5658 & 0 & 0 \\ 0 & 0 & 0.5658 & 0 \\ 0 & 0 & 0 & 0.5658 \end{pmatrix}, \qquad (6.17)$$

which results in an aligned system, where only one transmitter lies within the FOV of a receiver and vice versa. As a result, four completely independent links are formed. Fig. 6.5 displays the BER results of the three schemes for spectral efficiencies of SE = 4 and SE = 8 bits/s/Hz in this setup. In comparison to the $d_{\text{Tx}} = 0.6$ m scenario, all MIMO schemes perform worse because there is a lower received optical power. This is due to the missing cross-connects between emitter n_{t} and receiver n_{r} for $n_{\text{t}} \neq n_{\text{r}}$,

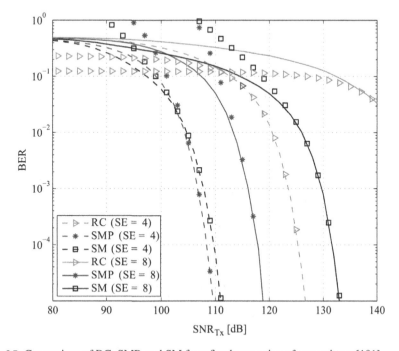

Figure 6.5 Comparison of RC, SMP, and SM for a fixed separation of transmitters [191]. Spectral efficiencies of SE = 4 bits/s/Hz and SE = 8 bits/s/Hz are considered in the 4 × 4 setup with $d_{\text{Tx}} = 0.7$ m (lines show simulation results and markers are analytical error bounds). ©2013 IEEE

as well as a larger path loss, leading to $\Delta^{\mathrm{SNR_{Rx}}}_{d_{\mathrm{Tx}}=0.7\mathrm{m}} \approx -16.95$ dB. While SM and SMP undergo a minor performance decrease of only about 3 dB, the performance of RC is degraded by about 10 dB. Consequently, RC is significantly penalized by the direct alignment, while SMP and SM can compensate the lower received optical power by the reduced channel correlation. For SE = 4 bits/s/Hz, SM again outperforms SMP in the low-$\mathrm{SNR_{Tx}}$ region up to about 103 dB. At higher $\mathrm{SNR_{Tx}}$, SMP outperforms SM due to its larger multiplexing gain. This issue is even more pronounced when SE = 8 bits/s/Hz is considered. In this case, SMP achieves major performance gains as it provides the spectral efficiency target with a lower modulation order of $M = 4$, while SM needs to operate with a higher modulation order of $M = 64$.

In the following, SM and the source of its performance gains, especially in the low-$\mathrm{SNR_{Tx}}$ region, are discussed in further detail. For this purpose, the BER of SM is segmented into errors arising from transmitter misdetection and from signal misdetection. If SE = 4 bits/s/Hz, SM conveys the same number of bits in the spatial and in the signal domain. Fig. 6.6 shows that for $d_{\mathrm{Tx}} = 0.2$ m, the errors caused by inaccurate detection of the transmitter have a significant impact on the BER, while the signal detection provides much lower error ratios. If $d_{\mathrm{Tx}} = 0.6$ m, transmitter detection provides a lower error ratio up to an $\mathrm{SNR_{Tx}}$ of about 99 dB, thus improving the overall BER of SM. At higher $\mathrm{SNR_{Tx}}$, signal detection can be performed more reliably and the errors caused by a misdetection of the transmitter again become the dominant factor. In the aligned

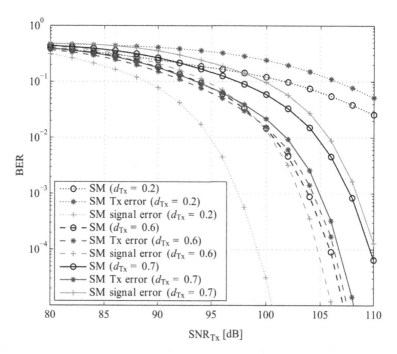

Figure 6.6 BER of SM segmented into bit errors caused by transmitter misdetection and signal misdetection [191]. Spectral efficiency of SE = 4 bits/s/Hz is considered in a 4 × 4 setup with varying distance of transmitters, d_{Tx}, on the X- and Y-axis. ©2013 IEEE

system with $d_{Tx} = 0.7$ m, the signal misdetection is the dominating source of errors due to the lower received optical power. In contrast, the detection of the active emitter can be performed more reliably because the direct alignment provides the lowest channel correlation. Therefore, the main property of SM to convey information in the spatial domain can be exploited most distinctively by direct alignment of the optical transmitters and receivers. Despite the lower received optical power, independent links enable the most reliable detection of the active transmitter.

6.4.2   Varying the position of receivers

In this section, the position of the receiver array is varied. Here, x_{Rx} and y_{Rx} are defined as the position offsets of the receiver array on the X- and Y-axis relative to the center of the room. These offsets increase both the distance and the misalignment between the transmitter array and the receivers. In order to still be able to detect the transmitter beams, a larger FOV semi-angle of the receivers of $\theta_{FOV,Rx} = 45°$ is assumed. The considered position offsets are chosen as follows: (i) $x_{Rx} = 0.5$ m, $y_{Rx} = 0$ m; (ii) $x_{Rx} = 0.25$ m, $y_{Rx} = 0.75$ m, and (iii) $x_{Rx} = 0.5$ m, $y_{Rx} = 1$ m. Applying (6.3) to these scenarios results in the following channel matrices:

$$
\mathbf{H_i} \approx 10^{-4}
\begin{pmatrix}
0.2293 & 0.2013 & 0.1462 & 0.1290 \\
0.2013 & 0.2293 & 0.1290 & 0.1462 \\
0.7964 & 0.6888 & 0.6410 & 0.5559 \\
0.6888 & 0.7964 & 0.5559 & 0.6410
\end{pmatrix},
$$

$$
\mathbf{H_{ii}} \approx 10^{-4}
\begin{pmatrix}
0.0461 & 0.0272 & 0.0358 & 0.0213 \\
0.2573 & 0.1798 & 0.1917 & 0.1352 \\
0.0735 & 0.0424 & 0.0713 & 0.0412 \\
0.4426 & 0.3040 & 0.4275 & 0.2940
\end{pmatrix},
$$

$$
\mathbf{H_{iii}} \approx 10^{-4}
\begin{pmatrix}
0.0061 & 0.0035 & 0.0044 & 0.0025 \\
0.0442 & 0.0283 & 0.0299 & 0.0194 \\
0.0150 & 0.0082 & 0.0128 & 0.0071 \\
0.1290 & 0.0792 & 0.1071 & 0.0663
\end{pmatrix}.
$$

(6.18)

Fig. 6.7 shows the BER of RC, SMP, and SM for a spectral efficiency of 8 bits/s/Hz in the 4×4 setup with $d_{Tx} = 0.4$ m using these position offsets. The increased distance between the transmitters and the receivers leads to a larger path loss and an increased SNR penalty as compared to the scenario with $x_{Rx} = y_{Rx} = 0$ m. However, the larger distance also increases the differences between the channels. This improves the performance of SMP. While in some cases the performance of SMP is degraded due to channel similarities, when the position offsets are zero, it now performs better than RC and SM, when position offsets and larger distances are considered. This is because of favorable channel conditions and the fact that the spatial multiplexing gain of SMP increases linearly with the minimum number of transmitters and receivers. In contrast, the spatial multiplexing gain of SM increases only logarithmically, and there is no spatial multiplexing gain in the case of RC.

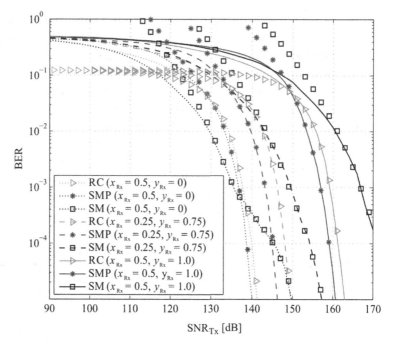

Figure 6.7 Comparison of RC, SMP, and SM with different position offsets of receiver array on the X- and Y-axis [191]. Spectral efficiency of SE = 8 bits/s/Hz is considered in a 4 × 4 setup with $d_{Tx} = 0.4$ m (lines show simulation results and markers are analytical error bounds). ©2013 IEEE

6.4.3   Power imbalance between transmitters

In the following, the effect of induced power imbalance between the individual transmitters is analyzed. For this purpose, the transmission power is not uniformly distributed across all N_t transmitters. The optical power imbalance factor is defined as δ, and it denotes the optical power surplus factor assigned to one transmitter in comparison to another. Note that the mean optical power emitted by all transmitters, $P_{s(opt)}$, is still the same as before. This means that the total transmission power is not increased by driving individual LEDs in the array with different powers. Moreover, the power distribution is done without any channel state information at the transmitter side. The optical power scaling factors for every transmitter to meet a given power imbalance of δ can be calculated as follows:

$$\alpha_1 = \frac{N_t}{\sum_{n_t=1}^{N_t} \delta^{n_t-1}},$$

$$\alpha_{n_t} = \delta \, \alpha_{n_t-1} .$$

(6.19)

For instance, if $\delta = 3$ dB and $N_t = 4$ are assumed, then $\alpha_1 \approx 4/15$, $\alpha_2 \approx 8/15$, $\alpha_3 \approx 16/15$, and $\alpha_4 \approx 32/15$. Using these factors, the optical transmission power assigned to emitter n_t employing RC or SMP is $\widetilde{P}_{s(opt),n_t} = P_{s(opt)}\alpha_{n_t}/N_t$, while for SM it is $\widetilde{P}_{s(opt),n_t} = P_{s(opt)}\alpha_{n_t}$. Note that the signal modulation technique is PAM according to (6.5) and (6.11), where $\widetilde{P}_{s(opt),n_t}$ is now the mean optical power assigned to transmitter n_t. In general, the receiver does not need any knowledge of the induced power imbalance because it implicitly obtains this information by channel estimation, which needs to be performed in all cases. Therefore, at the receiver, the power imbalance appears as a modified channel gain, and it does not increase the computational complexity of the signal detection.

Fig. 6.8 depicts the BER of RC, SMP, and SM for a spectral efficiency of 4 bits/s/Hz in the 4×4 scenarios with $d_{Tx} = 0.2$ m and $d_{Tx} = 0.6$ m for different power imbalances. As shown, adding an imbalance to the transmission powers can enhance the performance of both SMP and SM, while it has no influence on RC. This is because the correlation of the single links is reduced, making them more distinguishable at the receiver side. The performance of RC is not related to link differences but only to the absolute channel gains. Therefore, power imbalance has no effect on the performance of RC given the symmetrical arrangement of the channel gains denoted in (6.14). The results for the setup with $d_{Tx} = 0.2$ m shown in Fig. 6.8(a) indicate that a power imbalance of about $\delta = 1$ dB results in the best BER performance for SM, while $\delta = 3$ dB yields the lowest BER for SMP. It is also shown that higher power imbalances lead to worse error ratios. While the correlation of the links may be reduced, the transmission power for some of the links is largely decreased. This leads to a low SNR on these links and consequently to a worse overall BER performance. Therefore, a compromise between channel correlation and received SNR is required. This is also the reason why SMP can operate with higher power imbalances as compared to SM. Because of the fact that SM needs to operate with a larger signal constellation size in order to provide the same data rate, it is more susceptible to low SNRs. Consequently, SMP can benefit to a larger extent from the power imbalance, and it achieves the same performance as RC for $\delta = 3$ dB. Note that without power imbalance, SMP performs significantly worse than RC and SM in the setup with $d_{Tx} = 0.2$ m as shown in Fig. 6.4(a) because of the high channel correlation. For $d_{Tx} = 0.6$ m, power imbalance has a negative effect on SMP and SM, since their performance decreases with increasing δ as presented in Fig. 6.8(b). Thus, no further benefits can be achieved, and $\delta = 0$ dB gives the best performance for both SMP and SM.

6.4.4   Link blockage

In this section, the $d_{Tx} = 0.2$ m and $d_{Tx} = 0.4$ m scenarios are considered with an induced link blockage between some transmitters and receivers. This can be achieved by installing opaque boundaries in the receiver device, by smaller FOVs of the optical receivers, or by polarization filters. The assumed channel coefficients for the two scenarios are given in (6.14). However, some of the links of the 4×4 setup are intentionally blocked. These are the same 4 links which happen to have a zero path gain in the $d_{Tx} = 0.6$ m scenario, i.e. the links between Tx 1 and Rx 4, Tx 2 and Rx 3, Tx 3

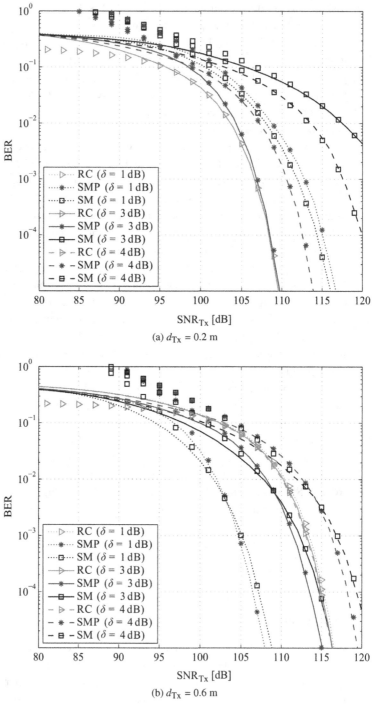

(a) $d_{Tx} = 0.2$ m

(b) $d_{Tx} = 0.6$ m

Figure 6.8 Comparison of RC, SMP, and SM with different power imbalances, δ [191]. Spectral efficiency of SE = 4 bits/s/Hz is considered in a 4×4 setup (lines show simulation results and markers are analytical error bounds). ©2013 IEEE

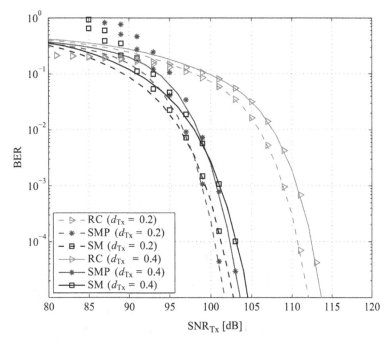

Figure 6.9 Comparison of RC, SMP, and SM with induced link blockage [191]. Spectral efficiency of SE = 4 bits/s/Hz is considered in a 4 × 4 setup with $d_{Tx} = 0.2$ m and $d_{Tx} = 0.4$ m (lines show simulation results and markers are analytical error bounds). ©2013 IEEE

and Rx 2, Tx 4 and Rx 1. This results in the following channel matrices for the two setups with induced link blockage:

$$\widehat{\mathbf{H}}_{d_{Tx}=0.2m} \approx 10^{-4} \begin{pmatrix} 1.0708 & 0.9937 & 0.9937 & 0 \\ 0.9937 & 1.0708 & 0 & 0.9937 \\ 0.9937 & 0 & 1.0708 & 0.9937 \\ 0 & 0.9937 & 0.9937 & 1.0708 \end{pmatrix}, \quad (6.20)$$

resulting in $\widehat{\Delta}_{d_{Tx}=0.2m}^{SNR_{Rx}} \approx -2.29$ dB, and

$$\widehat{\mathbf{H}}_{d_{Tx}=0.4m} \approx 10^{-4} \begin{pmatrix} 0.9226 & 0.7964 & 0.7964 & 0 \\ 0.7964 & 0.9226 & 0 & 0.7964 \\ 0.7964 & 0 & 0.9226 & 0.7964 \\ 0 & 0.7964 & 0.7964 & 0.9226 \end{pmatrix}, \quad (6.21)$$

with $\widehat{\Delta}_{d_{Tx}=0.4m}^{SNR_{Rx}} \approx -3.99$ dB.

Fig. 6.9 presents the BER of RC, SMP, and SM for SE = 4 bits/s/Hz in these two setups. Relative to the results without link blockage, RC performs about 2 dB worse. This is because the induced link blockage leads to a lower SNR at the receiver side. However, the performance of SM and SMP is significantly enhanced. Although there is a lower received optical power, both MIMO schemes benefit from reduced channel correlation. This is especially evident for SMP. Although SMP performs worst in this scenario without link blockage, it benefits from the reduced channel correlation in the setup with link blockage, and it provides the lowest error ratio. The setup with $d_{Tx} = 0.2$ m and induced link blockage provides the best compromise between channel correlation and received optical power. As shown, both SM and SMP perform about 2 dB better as compared to the setup with $d_{Tx} = 0.4$ m and induced link blockage. Relative to the setup with $d_{Tx} = 0.6$ m from Fig. 6.4(a), SM and SMP achieve an even larger performance gain of about 5 dB.

6.5   Summary

In this chapter, the performance of three MIMO techniques, RC, SM, and SMP, has been studied for OWC in indoor environments. Several 4×4 scenarios with different spacings of the transmitters and different positions of the receiver array have been considered. It has been shown that the MIMO schemes for OWC were able to provide gains even under static LOS conditions, i.e. channel conditions which generally render the use of MIMO difficult in the RF domain. SMP proved to be capable of increasing the spectrum efficiency of OWC with IM/DD transmission. In order to achieve these improvements, sufficiently low channel correlation was required. Similarly, SM achieved improved spectral efficiency especially in the low-SNR region, and it was more robust to high channel correlation. Furthermore, SM enjoyed additional implementation advantages, as it only required low complexity detection algorithms. This is because SM avoids ISI between the different spatial channels. Lastly, RC proved to be insensitive to different transmitter–receiver alignments, but it had the disadvantage that it required large signal constellation sizes to provide high data rates.

Several techniques that reduced channel correlation have been studied. It has been demonstrated that practical OWC systems were able to benefit greatly from adaptive MIMO techniques. These included variation of the spacing between transmitters and the relative position of the receiver array, induced power imbalance between the transmitters, and link blockage. It has been found that induced power imbalance between the transmitters was an effective technique to improve BER performance. Under conditions which caused high channel correlation, power imbalance enhanced the performance of both SMP and SM remarkably, as they best capitalized on both SNR and channel differences. Consequently, SMP and SM were applicable in scenarios which are generally considered as unfavorable for MIMO transmission. For the 4×4 indoor scenario, it has been shown that the best performance was achieved by blocking of some of the 16 links between the transmitters and receivers. The induced blockage reduced the SNR at the receiver side. However, the blocking of 4 links, for example, improved the BER of both SMP and SM by reducing channel correlation. As a result, this technique provided an effective SNR gain which outweighed the loss in SNR due to the lower received optical power. Therefore, the induced link blockage has proven to be the most suitable compromise between channel correlation and received optical power for the considered scenarios.

7 Throughput of cellular OWC networks

7.1 Introduction

An optical wireless communication (OWC) multiple access network can deliver wireless data to multiple mobile users. In order to maximize the coverage area, a cellular network approach can be deployed. The time and frequency resources can be allocated to the cells in a static fashion using resource partitioning with wavelength/frequency reuse or by means of a dynamic interference-aware scheduling. An aircraft cabin has been considered as a representative indoor scenario since: (a) it provides a very dense distribution of users, and hence results in an interference-limited environment; (b) it is a scenario where radio frequency (RF) communication is not preferred; and (c) it allows very small cells. It is well known that the introduction of small cells is one of the key reasons for the significant network spectral efficiency increase in RF cellular communications.

In this chapter, first, a cellular OWC system based on orthogonal frequency division multiplexing (OFDM), i.e. optical OFDM (O-OFDM), with static resource partitioning is considered inside an aircraft cabin. Asymmetrically clipped optical OFDM (ACO-OFDM) and direct-current-biased optical OFDM (DCO-OFDM) transmission schemes are compared in terms of throughput and cell coverage. Wavelength reuse factors of 1 (full reuse), 2, and 3 are studied, and off-the-shelf optical front-end components are assumed. The signal and interference distributions in an indoor cellular network can be accurately modeled by means of a Monte Carlo ray-tracing (MCRT) global irradiation simulation in a computer-aided design (CAD) cabin model. Due to the use of OFDM, adaptive modulation and coding (AMC) can be applied to the signal-to-interference-and-noise ratio (SINR) maps to obtain the spatial throughput distribution inside the cabin [16, 17]. It is shown that throughput within the range of 1.56−30 Mbps is achievable with the considered ACO-OFDM system, while DCO-OFDM doubles the throughput to 3.13−60 Mbps. In addition, wavelength reuse factors of 3 or higher are needed to ensure full cell support for both transmission schemes.

In the second section of this chapter, self-organizing interference management for optical wireless networks is studied [18]. A user that has received data in a given frame and intends to continue receiving data in the next frame broadcasts a busy burst (BB) in a time-multiplexed BB slot. The tagged access point (AP) intending to reuse a resource reserved in a neighboring cell must listen to the BB slot. Provided that channel reciprocity holds, the tagged AP infers prior to transmission the amount of co-channel interference (CCI) potentially caused towards the victim user in the neighboring cell. This is vital information for an AP to decide without any central supervision whether to transmit or defer the transmission to another time or frequency slot so as to limit CCI caused to the active link to a maximum value, defined as a threshold. The impact of this threshold parameter on the performance of optical orthogonal frequency division multiple access (OFDMA) networks deployed in an aircraft cabin is investigated. Simulation results demonstrate that the BB approach significantly improves both spectral efficiency and fairness in the system, compared to a static resource partitioning approach.

7.2   System throughput using static resource partitioning

Although the overall available bandwidth in the visible or infrared (IR) spectrum is in the THz range, the bandwidth of the signal that can be utilized by an OWC system with intensity modulation/direct detection (IM/DD) is inherently limited by the bandwidths of the transmitter and receiver front-ends. In order to serve multiple users and have ubiquitous system coverage, it becomes necessary to reuse the available bandwidth. Although bandwidth reuse potentially increases system capacity, transmission of data intended for a user inevitably causes interference to other users receiving data on the same resource and vice versa. Several interference coordination mechanisms have been investigated in the literature. Static resource partitioning [16, 63] using traditional cluster-based wavelength/frequency planning is the most commonly used approach. With this approach, the users served by a tagged AP are restricted to use a certain fixed subset of the available resources in the system. CCI is mitigated by ensuring that any two cells that reuse the same set of resources are separated in space by a minimum reuse distance.

In general, when frequency reuse is employed in conjunction with OFDMA, the modulation bandwidth of the optical front-ends is divided according to a frequency reuse factor. The resulting portions of the bandwidth are allocated to the cells such that the distance between the cells reusing the same bandwidth portion is maximized. This approach results in an increase of the SINR at the cell edge at the expense of a reduction of the available system bandwidth by the frequency reuse factor. Alternatively, static resource partitioning according to wavelength reuse can be employed for OWC. Here, a tagged cell can use a portion of the optical band in a fashion similar to wavelength division multiplexing (WDM) [10, 208]. In visible light communications (VLC), this can be achieved by the reuse of light emitting diodes (LEDs) with different colors, e.g. red, green, and blue. In the case of OWC in the IR spectrum, LEDs with different center wavelengths are reused. In both cases, optical filters are employed at the associated photodiodes (PDs). Therefore, wavelength reuse facilitates parallel optical channels employing the full modulation bandwidth of the optical front-ends. As a result, wavelength reuse retains the SINR enhancement property of frequency reuse without reduction of the modulation bandwidth. Consequently, the available system bandwidth and the share of bandwidth per user can be enhanced. This approach increases the cost of the receiver unit because separate filters and PDs with peak spectral response for each color/wavelength band are required.

An entirely different approach to enhance system capacity and link throughput is to use imaging optical concentrators, which can separate the optical signals impinging from different sources [209, 210]. The optical signals from two different sources would excite different regions of an imaging concentrator. As such, the data from different streams can be selected or rejected independently. However, such an approach would reduce the effective receiver area for the intended signal and therefore reduce the received signal-to-noise ratio (SNR). Furthermore, the cost of the imaging concentrators would be significantly higher than a standard PD used in optical receivers.

In this section, a cellular deployment of an OFDM-based OWC system with wavelength reuse is investigated. The optical signal-to-interference ratio (SIR) maps from Chapter 2 can be converted into electrical SINR maps [17, 33]. After optical-to-electrical (O/E) conversion of the intended signal and the interfering signal at a receiver, and also including the thermal and shot noise components, the electrical SINR maps are obtained for wavelength reuse factors of 1, 2, and 3. Practical center wavelengths of $\lambda_1 = 830$ nm, $\lambda_2 = 870$ nm, and $\lambda_3 = 940$ nm are assumed in combination with an optical bandpass filter at the intended receiver. In this way, adjacent channel interference (ACI) is avoided, whereas only CCI is present. Because of its inherent robustness to multipath fading and its capability of achieving high transmission rates, an OFDM-based system is considered. In the literature, two distinct system realizations can be found: ACO-OFDM [51, 138] and DCO-OFDM [147]. It has been reported that ACO-OFDM achieves a higher power efficiency in the low-SNR regime at the expense of 50% reduction in spectral efficiency as compared to DCO-OFDM in the high-SNR regime. In order to translate the electrical SINR maps into throughput, an AMC scheme [211] is employed. Finally, the throughput performance and the cell coverage of ACO-OFDM and DCO-OFDM are compared for the three wavelength reuse configurations. It is shown that the cellular OWC network inside the aircraft cabin is interference-limited due to the small cell size employed to facilitate high-capacity with high link density. Therefore, wavelength reuse factors of 3 or higher or interference-aware scheduling are required to ensure full cabin coverage.

7.2.1   Signal-to-interference-and-noise ratio modeling

In this section, the same model of the aircraft cabin and the MCRT simulation method as in Section 2.6 is used for the calculation of the throughput. The topologies of optical SIR distribution from Figs. 2.20(a), 2.21(a), and 2.22(a) are subjected to O/E conversion. The omnidirectional transmitters and receivers in the setup of the cellular network are based on off-the-shelf components. These include LEDs, PDs, optical bandpass filters, and transimpedance amplifiers (TIAs) [94–96, 100]. The network configurations for different wavelength reuse factors are presented in Fig. 2.9 in Section 2.6. Wavelength reuse in a cellular network is suitable for OWC, when individual optical channels can be separated by means of LEDs and PDs with a matching narrow optical spectral response or by means of optical filters. This can be achieved in VLC through independent modulation of the red, green, and blue (RGB) components of the white LED and RGB filters at the receiver. In IR OWC, the LEDs inherently have narrow optical

Table 7.1 Front-end setup [94–96, 100].

Wavelength	$\lambda_1 = 830$ nm	$\lambda_2 = 870$ nm	$\lambda_3 = 940$ nm
Transmitter	$16\times$ TSHG8200	$16\times$ TSFF5210	$16\times$ VSMB2020X01
FOV of LED	$\pm10^\circ$	$\pm10^\circ$	$\pm12^\circ$
$P_{\text{opt,ACO}}$	$16\times$ 10 mW	$16\times$ 10 mW	$16\times$ 10 mW
$P_{\text{opt,DCO}}$	$16\times$ 25 mW	$16\times$ 25 mW	$16\times$ 25 mW
Receiver	$4\times$ S6967	$4\times$ S6967	$4\times$ S6967
FOV of PD	$\pm70^\circ$	$\pm70^\circ$	$\pm70^\circ$
Area of PD, A	$4\times$ 26.4 mm^2	$4\times$ 26.4 mm^2	$4\times$ 26.4 mm^2
Responsivity, S_{PD}	0.6 A/W	0.63 A/W	0.65 A/W
Capacitance	$4\times$ 50 pF	$4\times$ 50 pF	$4\times$ 50 pF
Bandwidth	25 MHz	25 MHz	25 MHz
Optical filter	$4\times$ FB830-10	$4\times$ FB870-10	$4\times$ FB940-10
Transmittance, T_{OF}	0.8	0.8	0.8
Bandwidth	±10 nm	±10 nm	±10 nm
TIA	$1\times$ AD8015	$1\times$ AD8015	$1\times$ AD8015
Gain, G_{TIA}	10 kΩ	10 kΩ	10 kΩ
R_{load}	50 Ω	50 Ω	50 Ω
σ^2_{AWGN}	-38 dBm	-38 dBm	-38 dBm
BER	10^{-7}	10^{-7}	10^{-7}

spectral responses at the optical center wavelengths in the near IR (NIR) spectrum. In this study, wavelength reuse up to a factor of 3 is facilitated via IR channels at $\lambda_1 = 830$ nm, $\lambda_2 = 870$ nm, and $\lambda_3 = 940$ nm. The front-end parameters used in the simulation are given in Table 7.1. In addition, in accord with the characteristics of the components in the receiver, i.e. PD and TIA, -38 dBm of electrical power is considered for the additive white Gaussian noise (AWGN) component, representing the thermal and shot noise at the receiver. Including the non-linear distortion noise, the CCI and the AWGN components, the electrical SINR per symbol at the receiver can be expressed as follows:

$$\text{SINR} = \frac{P_{\text{S,elec}}}{P_{\text{clip,elec}} + P_{\text{I,elec}} + P_{\text{N,elec}}}$$

$$= \frac{\left(P_{\text{S,opt}} g_{\text{S}}\right)^2 F_{\text{OE}} F_{\text{O,S}}^2 K^2 G_{\text{DC}} / G_{\text{B}}}{\sigma_{\text{clip}}^2 g_{\text{S}}^2 G_{\text{DC}} + \left(\sum_i P_{\text{I,opt}} g_{\text{I},i} \sqrt{F_{\text{OE,I},i}}\right)^2 / G_{\text{B}} + \sigma_{\text{AWGN}}^2 G_{\text{B}}}, \tag{7.1}$$

where the intended electrical symbol power, the non-linear distortion noise power, the total interference electrical power, and the electrical noise power are defined as $P_{\text{S,elec}}$, $P_{\text{clip,elec}}$, $P_{\text{I,elec}}$, and $P_{\text{N,elec}}$, respectively. The radiated average optical power per transmitter in the ACO-OFDM and DCO-OFDM systems is given in Table 7.1 as $P_{\text{opt,ACO}}$ and $P_{\text{opt,DCO}}$, respectively. Here, the equal average optical power at the intended and interfering transmitters, $P_{\text{S,opt}}$ and $P_{\text{I,opt}}$, is considered. Due to the statistics of the signal elaborated in Section 4.4.1, the ACO-OFDM system radiates less optical power than the DCO-OFDM system. Here, the factor g_S denotes the optical path gain between the intended transmitter and the intended receiver, while $g_{I,i}$ defines the optical path gain between the ith interfering transmitter and the intended receiver. The optical path gains can be obtained from (2.16) in Section 2.5.2. The path gains are dependent on the area of the PD, A, as elaborated throughout the derivations of the received optical power in Section 2.5.2, and they are calculated from the irradiance distribution in the cabin. The additional parameters for the calculation of the path gains, such as the responsivity of the PD, S_{PD}, the gain of the TIA, G_{TIA}, the resistive load over which the received signal is measured, R_{load}, and the transmittance of the optical filter, T_{OF}, are listed in Table 7.1. An optical concentrator is not considered in this setup. The attenuation factor of the non-linear distortion, K, and the variance of the non-linear distortion noise, σ_{clip}^2, are derived in Chapter 3. Both in ACO-OFDM and DCO-OFDM, the attenuation factor, K, is given in (3.8) and (3.19) for a generalized distortion function and for double-sided signal clipping, respectively. In ACO-OFDM, the non-linear distortion noise variance, σ_{clip}^2, is given in (3.18) and (3.24), while in DCO-OFDM it is given in (3.12) and (3.22) for the two considered distortion functions. The electrical power penalty due to the direct-current (DC) bias, G_{DC}, can be obtained from (4.21) and (4.20) in Section 4.4.1 for ACO-OFDM and DCO-OFDM, respectively. The bandwidth utilization factor, G_B, is given as 0.5 in ACO-OFDM and $(N-2)/N$ in DCO-OFDM, which approaches unity for a large number of subcarriers, N. The O/E conversion factor for the undistorted optical and electrical symbol powers, F_{OE}, can be obtained in ACO-OFDM and DCO-OFDM from (4.17) and (4.16), respectively, as follows: $F_{OE} = P_{s(elec)}/P_{s(opt)}^2$. In a front-end setup with strong non-linear distortion, the average radiated optical power deviates from the undistorted optical symbol power. Therefore, in order to reveal how much useful optical symbol power, and therefore useful electrical symbol power via F_{OE}, corresponds to a given average received optical power level, $E[F(x)]$, a further transformation factor for the intended signal, $F_{O,S}$, is required, and it is defined as follows: $P_{s(opt)}/E[F(x)]$. Here, $E[F(x)]$ can be expressed as $E[F(x)] = E[\Xi(x)] = E[\Psi(s)] + \Xi(\beta_{DC})$ for the generalized distortion function $F(x) = \Xi(x)$, where β_{DC} is the DC bias, and $E[\Psi(s)]$ can be obtained from (3.11) and (3.12) for both ACO-OFDM and DCO-OFDM. For the case of double-sided signal clipping, $E[F(x)]$ can be expressed as $E[F(x)] = E[\Phi(x)] = E[\Psi(s)] + \Phi(\beta_{DC})$ for the double-sided clipping function $F(x) = \Phi(x)$ from (3.2). Here, $E[\Psi(s)]$ can be obtained from (3.21) for both ACO-OFDM and DCO-OFDM. In addition, in the latter case, $E[F(x)] = E[\Phi(\mathbf{x}_l)]$ can also be found in (4.15). Since the signal and the non-linear distortion noise from the interfering transmitters cannot be separated at an intended transmitter, they both are perceived as interference. Therefore, in order to determine how much interfering electrical power corresponds to interfering average received optical power, a transformation factor is defined for the individual interfering optical signals, $F_{OE,I,i}$. This is because there exists the possibility that every interfering AP is setup with a different signal standard deviation, σ, and DC bias, β_{DC}, within its dynamic range. In ACO-OFDM, $F_{OE,I,i}$ is defined as $F_{OE,I,i} = E\left[F(x - P_{min,norm})^2\right]/E[F(x)]^2$ according to the derivation in (3.15) and (3.16). For a generalized distortion function, $F_{OE,I,i} = E[\Psi(s - \sigma\lambda_1)^2]/E[\Xi(x)]^2$, where $E[\Psi(s - \sigma\lambda_1)^2]$ can be obtained from (3.16) and (3.18), while $E[\Xi(x)]^2$ is the same as above. In the case of double-sided signal clipping, $F_{OE,I,i} = E[\Psi(s - \sigma\lambda_{bottom})^2]/E[\Phi(x)]^2$, where $E[\Psi(s - \sigma\lambda_{bottom})^2]$ can be obtained from (3.23), while $E[\Phi(x)]^2$ is the same as above. In DCO-OFDM, $F_{OE,I,i}$ is defined as $F_{OE,I,i} = (E[F(x)^2] - E[F(x)]^2)/E[F(x)]^2 = (E[\Psi(s)^2] - E[\Psi(s)]^2)/E[F(x)]^2$. Here, $E[\Psi(s)^2]$ and $E[\Psi(s)]^2$ can be obtained from (3.11) and (3.12) for a generalized non-linear distortion function, as well as (3.20) and (3.21) for double-sided signal clipping, while $E[F(x)]^2$ is the same as above.

In this study, a positive infinite linear dynamic range of the transmitter with $P_{min,norm} = 0$ is considered for simplicity. Considering a DC bias of $\beta_{DC} = 0$ in ACO-OFDM and $\beta_{DC} = \kappa\sigma$ in DCO-OFDM, F_{OE} is given as π and $(\kappa^2 + 1)/\kappa^2$, respectively. In this study, $\kappa = 3$ is considered in DCO-OFDM, which yields $F_{OE} = 9/10$ and a bottom-level clipping of $\lambda_{bottom} = -3$. As a result, negligible signal clipping distortion can be considered for both O-OFDM systems. Therefore, $K = 0.5$ in ACO-OFDM and $K = 1$ in DCO-OFDM, as well as $\sigma_{clip}^2 = 0$, can be considered. In addition, $G_{DC} = 1$ and $G_{DC} = 1/(\kappa^2 + 1) = 1/10$ are obtained for ACO-OFDM and DCO-OFDM, respectively. Since the non-linear distortion in this setup is negligible, $F_{O,S} = 1$ and $F_{OE} = F_{OE,I,i}$.

The received signal must be received with a certain minimum electrical SINR which is sufficient to decode the signal according to the modulation and forward error correction (FEC) coding format used for transmission with a bit-error ratio (BER) lower than a tolerated limit. The maximum tolerable BER figure depends on the application, but in general it is considered to be 10^{-3} for voice and 10^{-7} for data. The transmitter and receiver pair can communicate with each other as long as the minimum SINR target needed for decoding the lowest order modulation and coding scheme with the required quality of service (QoS) is met. In this study, a practical BER of less than 10^{-7} is targeted for video broadcast.

7.2.2   Adaptive modulation and coding

The obtained electrical SINR maps are transformed into throughput maps by means of an AMC scheme. A combination of multi-level quadrature amplitude modulation (M-QAM), i.e. $M = \{4, 16, 64\}$, and various coding rates of a forward error correction (FEC) convolutional coding scheme, i.e. $R = \{4/5, 3/4, 2/3, 1/2, 1/3, 1/4, 1/5, 1/8\}$, are used. As a result, a wide operational SINR range is facilitated. The performance of an RF OFDM communication system for the chosen AMC scheme is given in [211]. Accordingly, ACO-OFDM achieves a quarter of the throughput of the RF OFDM system for a given BER/SINR target, while DCO-OFDM delivers half the throughput. Therefore, the SINR requirements for the 10^{-7} BER target and the corresponding throughput of RF OFDM with AMC from [211] can be used to obtain the throughput of ACO-OFDM and DCO-OFDM given in Table 7.2. As a result, the ACO-OFDM system switches transmission rates between 1.56 Mbps and 30 Mbps over a range of electrical SINR values between -5.1 dB and 17.8 dB. The more spectrally efficient DCO-OFDM system delivers double throughput, i.e. $3.13-60$ Mbps. In the next section, the optical

Table 7.2 SINR target and throughput of ACO-OFDM and DCO-OFDM for M-QAM and convolutional codes with a coding rate R

AMC scheme	SINR target at 10^{-7} BER [dB]	Throughput [Mbps]	
		ACO-OFDM	DCO-OFDM
4-QAM, $R = 1/8$	−5.1	1.56	3.13
4-QAM, $R = 1/5$	−3	2.5	5
4-QAM, $R = 1/4$	−1.9	3.13	6.25
4-QAM, $R = 1/3$	−0.2	4.17	8.33
4-QAM, $R = 1/2$	2	6.25	12.5
4-QAM, $R = 2/3$	4.1	8.33	16.67
4-QAM, $R = 4/5$	6.2	10	20
16-QAM, $R = 1/2$	7.4	12.5	25
16-QAM, $R = 2/3$	11.1	16.67	33.33
16-QAM, $R = 4/5$	12.4	20	40
64-QAM, $R = 2/3$	14.7	25	50
64-QAM, $R = 3/4$	16.6	28.13	56.25
64-QAM, $R = 4/5$	17.8	30	60

SIR maps from Chapter 2 are converted into effective electrical SINR maps for the two O-OFDM schemes through (7.1), which in turn are converted into throughput distribution inside the aircraft cabin by means of Table 7.2.

7.2.3   System throughput of optical OFDM in an aircraft cabin

The throughput distribution in the cabin is presented in Figs. 7.1 and 7.2 for ACO-OFDM and DCO-OFDM, respectively. The key results are summarized in Table 7.3. The OWC systems are compared in terms of the delivered throughput range and the corresponding cell coverage. The results show that the underlying electrical SINR maps enable both O-OFDM systems to span their entire throughput range, i.e. ACO-OFDM demonstrates transmission rates between 1.56 Mbps and 30 Mbps, while DCO-OFDM doubles the rates to 3.13−60 Mbps. For wavelength reuse factors of 1 and 2, ACO-OFDM supports a larger portion of the cell area as compared to DCO-OFDM. It is shown that both systems require wavelength reuse factors of 3 or higher in order to ensure a full cell coverage. In the latter case, the "dead zone" in the considered communication scenario only appears in the areas which lie beyond the boundaries of the cabin geometry or inside opaque objects. In addition, Figs. 7.1(c) and 7.2(c) highlight the trade-off between power efficient and spectrally efficient systems. The more power efficient ACO-OFDM system delivers a higher throughput towards the cell edge, where the SINR is low. However, the more spectrally efficient DCO-OFDM doubles the throughput towards the cell center, where the SINR is high. As a result, the throughput advantage of the power efficient ACO-OFDM system in the lower SINR region becomes inferior to the more spectrally efficient DCO-OFDM system in the higher SINR region. This can be explained by the fact that ACO-OFDM requires an M^2-QAM modulation to achieve the same throughput as M-QAM DCO-OFDM. Therefore, the

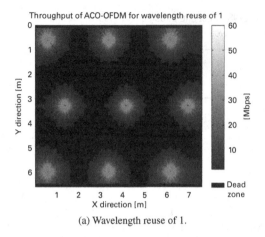

(a) Wavelength reuse of 1.

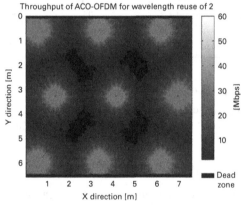

(b) Wavelength reuse of 2.

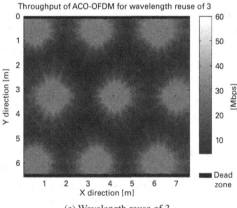

(c) Wavelength reuse of 3.

Figure 7.1 Throughput of ACO-OFDM for wavelength reuse factors of 1, 2, and 3

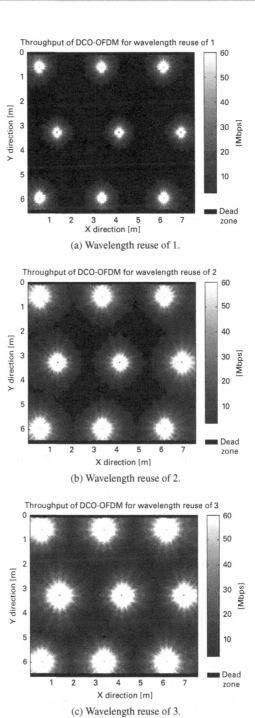

(a) Wavelength reuse of 1.

(b) Wavelength reuse of 2.

(c) Wavelength reuse of 3.

Figure 7.2 Throughput of DCO-OFDM for wavelength reuse factors of 1, 2, and 3

Table 7.3 Throughput in the cell at (4.15 m, 3.25 m) in the cross-section through the transmitters

OFDM scheme	ACO-OFDM			DCO-OFDM		
Wavelength reuse factor	1	2	3	1	2	3
Minimum throughput [Mbps]	1.56	1.56	4.17	3.13	3.13	3.13
Minimum throughput radius [m]	0.91	1.21	1.7	0.68	1.13	1.7
Maximum throughput [Mbps]	30	30	30	60	60	60
Maximum throughput radius [m]	0.23	0.42	0.61	0.15	0.3	0.49

ACO-OFDM system using squared order modulation requires a much higher SINR in order to achieve the BER target as compared to DCO-OFDM for the same high transmission rate. In general, a multi-user access scheme other than frequency division multiple access (FDMA) may consider an adaptive switching between ACO-OFDM and DCO-OFDM. For example, OFDM time-domain multiple access (OFDM-TDMA) can be employed. Such an approach enables the allocation of an optimum transmission rate according to the underlying SINR conditions in OFDM-based OWC systems.

7.3   Interference coordination in optical cells using busy burst signaling

If the instantaneous traffic loads vary widely among cells within the network [212], static resource partitioning can lead to wastage of resources in lightly loaded cells and failure to cater for the traffic demands in heavily loaded cells. Using O-OFDM as a transmission scheme, the available bandwidth served by an AP can be shared among multiple users by assigning each user a different amount of bandwidth corresponding to user demand and scheduling policies. Likewise, link adaptation can be carried out to scale the user throughput according to the prevalent channel conditions at the receiver. An OFDMA network, where the resource chunks (time–frequency slots) are fully reused, is prone to high CCI at the cell edge and because of this the throughput is particularly reduced for the cell-edge users. Therefore, interference coordination is essential among multiple optical cells within the network in order to balance system capacity against enhanced throughput at the cell edge. In order to enhance bandwidth reuse and mitigate interference in a cost-effective manner, cognitive spectrum sensing approaches will be needed. To this end, a classical carrier sense multiple access with collision detection (CSMA/CD) approach used in the Ethernet has been considered for optical wireless applications [213]. However, it is well known that the CSMA/CD approach has hidden node and exposed node problems, both of which degrade the performance in a wireless network.

To address the aforementioned challenges, interference-aware allocation of resource chunks using BB signaling for transmitting data in optical IM/DD-based OFDMA time division duplexing (OFDMA-TDD) systems can be employed. With the proposed approach, the assignment of chunks in a tagged AP is adjusted dynamically depending on the location of an active user in the neighboring cell. To facilitate this, each user equipment (UE) must broadcast a BB [128, 129] in a time-multiplexed slot after successfully receiving data in order to reserve the chunk for the next frame. The AP that intends to transmit on a given chunk must listen to the BB slot corresponding to that chunk. Provided that channel reciprocity [127] holds, the AP infers the amount of CCI it could potentially cause towards the user that has reserved the chunk. This vital information allows an AP to decide without any central supervision whether to transmit or defer the transmission to another time and/or frequency slot so as to limit the CCI caused to the active link to a threshold value. The impact of this threshold parameter on the performance of optical OFDMA networks deployed in an aircraft cabin is investigated. Extensive system-level simulations demonstrate that performing chunk allocation using the BB protocol enhances the mean system throughput by 17%, whilst maintaining the same throughput at the cell edge compared to that achieved with static resource partitioning. In addition, it is shown that when the offered load begins to exceed the traffic capacity, hardly any chunks available for a tagged AP are idle. As a result, a new user entering the network or an existing user switching from idle (empty buffer) to active (with a packet in the buffer) state would experience outage. A heuristic that annuls reservation after a user has had its fair share of resources is proposed. Simulation results show that the BB protocol combined together with the proposed heuristic significantly improves both the guaranteed user throughput and the median system throughput compared to the static chunk allocation using cluster-based resource partitioning.

7.3.1   System model

An optical wireless network where U users are served by N_A optical APs is considered for the downlink mode. The transceiver module consists of an LED (or an array of LEDs) and a PD (or an array of PDs) for transmitting and receiving optical signals, respectively. An OFDMA-TDD air interface is employed, where the available system bandwidth B is divided into N subcarriers. The available OFDMA subcarriers are grouped in contiguous blocks made up of n_{sc} subcarriers and n_{os} OFDM symbols. Such blocks form a resource unit called a chunk, denoted by (m_c, k_c), where m_c is the chunk frequency index, and k_c is the chunk time index. A detailed description of the physical layer in O-OFDM transmission is presented in Chapter 4. In order to ensure a real-valued time-domain signal, Hermitian symmetry is imposed on the subcarriers, while subcarriers with indices 0 and $N/2$ are set to zero as illustrated in Fig. 7.3. This restriction limits the number of chunks that can be independently allocated to half of what is available within the system bandwidth.

In OFDMA-TDD, bit loading is considered for the individual users according to an AMC scheme as discussed in Section 7.2.2 and further elaborated in Section 7.3.6.

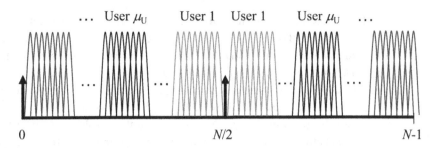

Figure 7.3 Assignment of subcarriers to users in optical OFDMA-TDD

Thus, the received information-carrying subcarrier in O-OFDM belonging to chunk m_c can be expressed from (4.19) in Chapter 4 as follows:

$$\widetilde{\mathbf{f}}_{\text{info}}^{m_c} = \left(K \mathbf{f}_{\text{info}}^{m_c} / \sqrt{G_B} + \mathbf{W}_{\text{clip}}^{m_c} \right) g_{\text{h(opt)}} \sqrt{G_{\text{DC}}} + \sqrt{G_B} \mathbf{W}_{m_c} , \qquad (7.2)$$

where K is the attenuation actor of the non-linear distortion, and $\mathbf{W}_{\text{clip}}^{m_c}$ is the zero-mean Gaussian non-linear distortion noise component with a variance of $\sigma_{\text{clip}}^2 n_{\text{sc}}/N$ presented in Chapter 3. The optical path gain of the flat fading optical wireless channel is denoted by $g_{\text{h(opt)}}$, and here it is considered as given in (6.3) from Chapter 6. The AWGN component is denoted as $\mathbf{W}_{m_c}$, and its variance is given by $\sigma_{\text{AWGN}}^2 n_{\text{sc}}/N$. This model is used below to obtain the components in the electrical SINR per chunk.

7.3.2   Interference coordination in optical cells

A simplified model of the considered optical wireless cellular system is depicted in Fig. 7.4. The observation angle of the transmitter in the AP and the incident angle of the receiver in the UE on the direct path in the downlink are given as $\theta_{\text{Tx,d}}$ and $\theta_{\text{Rx,d}}$, respectively. Here, the access point AP_1 is the intended transmitter for the receiver in user equipment UE_1. The access point AP_2 serving a different UE is an interfering transmitter for UE_1, when they are within the FOV.

For the analysis hereafter, let μ_U denote a UE which is associated with an AP α_A. Likewise, UE ν_U is another UE which is served by an AP β_A using the same chunk that is used by AP α_A to serve UE μ_U. Therefore, AP α_A causes CCI to UE ν_U, and AP β_A causes CCI to UE μ_U.

In order to distinguish between intended and interfering signals, subscripts of the form g_{α_A,μ_U} are added to the optical channel gains, where the first subscript denotes the transmitter and the second one denotes the receiver. From the perspective of UE μ_U, the desired and interfering electrical signal powers can be expressed from (7.1), respectively, as follows:

$$P_{\text{S,elec}}^{\mu_U} = \left(P_{\text{opt}}^{\alpha_A} g_{\alpha_A,\mu_U} \right)^2 F_{\text{OE}}^{\alpha_A} F_{\text{O},\alpha_A}^2 K_{\alpha_A}^2 G_{\text{DC}}^{\alpha_A} / G_B , \qquad (7.3)$$

$$P_{\text{I,elec}}^{\mu_U} = \left(P_{\text{opt}}^{\beta_A} g_{\beta_A,\mu_U} \right)^2 F_{\text{OE},\beta_A} / G_B . \qquad (7.4)$$

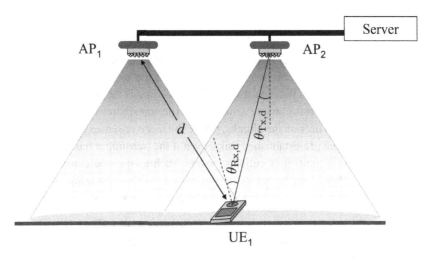

Figure 7.4 Illustration of interference in optical wireless network for the downlink. AP_1 is the intended transmitter for the receiver labeled UE_1 and AP_2 is an interfering transmitter which serves a different UE

Here, $P_{opt}^{\alpha_A}$ and $P_{opt}^{\beta_A}$ are the optical powers of APs α_A and β_A, respectively. The attenuation factor of the non-linear distortion at AP α_A is denoted as K_{α_A}, while $G_{DC}^{\alpha_A}$ is the DC-bias penalty. The O/E conversion factors for the undistorted optical power at AP α_A are given as $F_{OE}^{\alpha_A}$ and F_{O,α_A}, and they are given as defined in Section 7.2.1. The O/E factor for the interfering AP β_A is given as F_{OE,β_A}.

Since the subcarriers are assigned in chunks, the SINR is constant for all subcarriers within the chunk. The SINR at UE μ_U on chunk (m_c, k_c), including the non-linear distortion noise and the AWGN, is expressed as follows:

$$
\gamma_{\mu_U}[m_c, k_c] = \frac{P_{S,elec}^{\mu_U}[m_c, k_c]}{P_{clip,elec}^{\mu_U}[m_c, k_c] + P_{I,elec}^{\mu_U}[m_c, k_c] + P_{N,elec}^{\mu_U}[m_c, k_c]}
$$

$$
= \frac{\left(P_{opt}^{\alpha_A} g_{\alpha_A, \mu_U}\right)^2 F_{OE}^{\alpha_A} F_{0,\alpha_A}^2 K_{\alpha_A}^2 G_{DC}^{\alpha_A}/G_B}{\sigma_{clip,\alpha_A}^2 g_{\alpha_A, \mu_U}^2 G_{DC}^{\alpha_A} n_{sc}/N + \left(P_{opt}^{\beta_A} g_{\beta_A, \mu_U}\right)^2 F_{OE,\beta_A}/G_B + \sigma_{AWGN,\mu_U}^2 G_B n_{sc}/N}.
$$

(7.5)

Here, σ_{clip,α_A}^2 is the variance of the non-linear distortion noise at AP α_A which can be obtained at UE μ_U as shown in Chapter 3 when knowing the front-end biasing setup of AP α_A. The variance of the AWGN at UE μ_U is denoted as σ_{AWGN,μ_U}^2. In order to generalize the SINR expression in (7.5) when all the interfering APs are active, the interference electrical power can be summed as shown in (7.1).

Interference coordination using BB signaling comprises the following key components:

1. Interference avoidance: The transmitter avoids causing detrimental CCI towards the receivers of pre-established links. This is achieved by sensing the BB signal to determine the idle chunks as described in Section 7.3.3.

2. Contention mitigation: The access of chunks that are not yet reserved by transmitting a BB signal is regulated such that at most one transmitter within a coordination cluster may access the idle chunks. Such coordination is required so as to ensure that no two transmitters sense the same chunk as idle and cause collision of data in the next frame. This procedure is described in Section 7.3.4.

3. Interference-aware scheduling: The transmitter does not schedule transmissions for a user on the chunks, where the receiver senses a high level of CCI originating from pre-established links, even if the transmitter has sensed the chunk to be idle. In a multi-user cellular system, such as the one considered in this chapter, the chunk can be reused at the tagged AP by allocating it to a different user that reports a lower *a priori* CCI level. This procedure is described in Section 7.3.5.

Provided that an AP is aware of the amount of CCI it causes to a user served by a neighboring cell, the AP can use this knowledge to determine whether or not it may utilize a given chunk for serving its own user. This knowledge can be obtained by utilizing receiver initiated feedback in the form of a BB as detailed in Section 7.3.3.

7.3.3 Busy burst principle

For interference coordination among optical wireless cells, the medium access control (MAC) frame is divided into data and BB slots. Time division duplexing (TDD) is assumed, where a base station and a UE use the same frequency channel for transmission and reception. The link directions are separated in time. Data transmission is carried out during the data slot, and the BB is transmitted during the BB mini slot. Each receiver must transmit a BB to announce its presence to other APs, so that they may autonomously determine whether or not they can reuse the reserved chunk. To this end, it is assumed that UE μ_U has transmitted a BB after receiving data during its data slot. The BB signal transmitted by UE μ_U propagates through the optical channel and is received at both APs α_A and β_A. The received signal at the serving AP α_A informs the transmitter that the minimum SINR target is met. Furthermore, additional information can be piggybacked onto the BB signal for feedback and control purposes. By contrast, the received BB signal strength at the AP β_A is proportional to the amount of CCI it potentially causes to the active receiver, μ_U, if it were to reuse the reserved resource. By measuring the signal power received during the BB slot, the AP β_A decides autonomously whether or not it may reuse the reserved chunk so as to limit the interference to a threshold value. To this end, received electrical power of the BB signal from UE μ_U at AP β_A is given as follows:

$$P_{\text{bb,elec}}^{\beta_A} = \left(P_{\text{opt}}^{\text{bb}} g_{\mu_U,\beta_A} \right)^2 F_{\text{OE},\mu_U}/G_B , \qquad (7.6)$$

where $P_{\text{opt}}^{\text{bb}}$ is the radiated optical power of the BB signal from UE μ_U. Combining (7.4) and (7.6), and provided that $g_{\beta_A,\mu_U} = g_{\mu_U,\beta_A}$ and $F_{\text{OE},\beta_A} = F_{\text{OE},\mu_U}$, $P_{\text{I,elec}}^{\mu_U}$ can be expressed in terms of $P_{\text{bb,elec}}^{\beta_A}$ as follows:

$$P_{\text{I,elec}}^{\mu_U} = \left(\frac{P_{\text{opt}}^{\beta_A}}{P_{\text{opt}}^{\text{bb}}}\right)^2 P_{\text{bb,elec}}^{\beta_A} . \tag{7.7}$$

The channel reciprocity condition $g_{\beta_A,\mu_U} = g_{\mu_U,\beta_A}$ easily holds true in a line-of-sight (LOS) scenario, where the transmitter and receiver in the AP/UE have the same orientation, as well as the same radiation and detection patterns. In addition, given that the same front-end biasing setup is applied to AP β_A and UE μ_U, the second condition $F_{\text{OE},\beta_A} = F_{\text{OE},\mu_U}$ is also true. Furthermore, if $P_{\text{opt}}^{\beta_A} = P_{\text{opt}}^{\text{bb}}$, from (7.7) it follows that:

$$P_{\text{I,elec}}^{\mu_U} = P_{\text{bb,elec}}^{\beta_A} . \tag{7.8}$$

The AP β_A may reuse the resource if $P_{\text{I,elec}}^{\mu_U} \leq P_{\text{I,elec,th}}$, where $P_{\text{I,elec,th}}$ is the threshold interference electrical power. Furthermore, provided that both the transmitted data electrical power and the BB electrical power are equal, the condition for reusing a resource is given as follows:

$$P_{\text{bb,elec}}^{\beta_A} \leq P_{\text{I,elec,th}} . \tag{7.9}$$

Hence, by measuring the BB power observed during the feedback slot, AP β_A determines whether or not it causes detrimental CCI (i.e. CCI higher than the pre-determined threshold value) towards the active user μ_U served in an adjacent cell α_A. For a given threshold $P_{\text{I,elec,th}}$, (7.9) is more likely to hold true if μ_U lies close to AP α_A compared to the case if μ_U lies at the boundary of the coverage regions of APs α_A and β_A. Therefore, each AP can dynamically determine the set of chunks it can use for transmission, whilst avoiding detrimental CCI to pre-established links, simply by measuring the received BB power and comparing it against the threshold value without requiring a central coordinator.

The BB protocol relies on the assumption that AP α_A and AP β_A do not check (7.9) simultaneously. If both AP α_A and AP β_A were to sense the channel simultaneously, they would both infer that the channel is free, since there is no pre-established receiver in the vicinity that announces its presence by emitting a BB signal. In such a scenario, they would both schedule transmission for their own users without any knowledge of the user the other AP might schedule. Hence, each of the APs can potentially cause severe CCI to the user served by the other AP. Such collision can potentially cause high outage, particularly at the cell edge. To mitigate such a problem, the cellular slot access and reservation (CESAR) approach [130] is used, which ensures that an idle chunk is not assigned simultaneously by neighboring APs through a cyclically shifted allocation pattern for idle chunks.

7.3.4 Contention avoidance among neighboring cells

The contention problem can be mitigated by exploiting the properties of a cellular network using the CESAR approach proposed in [130] and illustrated in Fig. 7.5. In summary, the available chunks within the system bandwidth are grouped into S different subbands. Likewise, the cells in the system are also grouped into S different groups, $\mathcal{G}$, such that the APs within a coordination cluster belong to different groups. For the

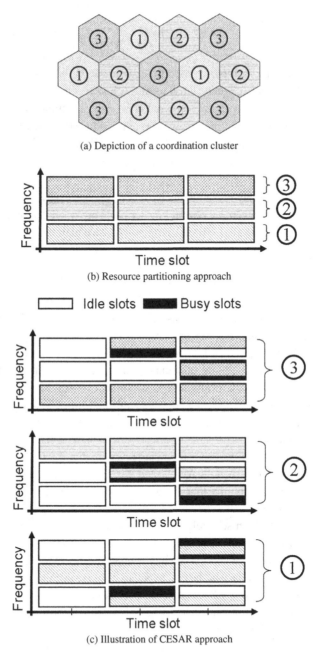

(a) Depiction of a coordination cluster

(b) Resource partitioning approach

(c) Illustration of CESAR approach

Figure 7.5 Collision avoidance using the CESAR approach [130]. The size of the coordination cluster (a) is set such that any two transmitter and receiver pairs that could cause collisions are within the coordination cluster, which is a principle followed for resource partitioning (b). Only one cell within a coordination cluster accesses idle chunks and avoids transmitting on those chunks, where a BB is sensed above the threshold value. The idle chunks that are successfully accessed are reserved and used for transmission in subsequent slots (c)

optical network, a reasonable choice of the number of groups is $S = 3$. At time instant k_c, the APs β_A in the group $\mathcal{G}_{\beta_A}[m_c, k_c] \in \mathcal{G}$ may access idle chunk (m_c, k_c) in a predefined cyclic manner determined as follows [130]:

$$\mathcal{G}_{\beta_A}[m_c, k_c] = \text{mod}(m_c + k_c, S) . \tag{7.10}$$

The term $\mathcal{G}_{\beta_A}[m_c, k_c] \in \mathcal{G}$ determines whether or not a chunk (m_c, k_c) may be accessed by AP β_A. Hence, (7.10) establishes a schedule to access idle chunks, i.e. chunks that are not protected by transmitting the BB. Thus, the CESAR mechanism ensures that at a particular time instant, k_c, only one group of APs may access idle resources, thereby mitigating the contention problem. Cyclically shifting the AP groups in (7.10) over time ensures that after S slots an AP is granted access to all chunks that are sensed idle. The APs are numbered such that an AP is associated to its group by the relation $\mathcal{G} = \{\text{mod}(\beta_A, S)\}$. Then, the AP β_A may allocate chunk (m_c, k_c) to user ν_U, only if both of the following conditions hold:

1. $\text{mod}(\beta_A, S) = \text{mod}(m_c + k_c, S)$ indicates that AP β_A may access chunk (m_c, k_c).
2. The threshold test (7.9) applied on chunk (m_c, k_c) holds true.

The second condition indicates that chunks already reserved by transmission of a BB signal retain unrestricted access to a given resource unit (m_c, k_c) and continue to serve the users that have reserved those chunks.

CESAR [130] and BB protocol for interference coordination in an optical wireless network perfectly complement each other. The former mitigates collisions due to simultaneous access (7.10), while the latter facilitates interference-aware selection of chunks reserved by transmitting the BB signal in order to limit the CCI caused to users served by the neighboring AP to a threshold value. However, UE ν_U also experiences CCI originating from pre-established links which make no effort to limit the CCI caused to the newly entering link. Hence, the minimum SINR target may not be met for the cell-edge user served by the tagged AP if the user is served using the chunks that are in use in the neighboring cell. To avoid scheduling UE ν_U on a chunk, where the CCI caused by pre-established links would cause the minimum SINR target not to be met, an *a priori* estimate of the electrical SINR, $\hat{\gamma}_{\nu_U}[m_c, k_c]$, is made as follows:

$$\hat{\gamma}_{\nu_U}[m_c, k_c] = \frac{P_{S,\text{elec}}^{\nu_U}[m_c, k_c]}{P_{\text{clip,elec}}^{\nu_U}[m_c, k_c] + P_{I,\text{elec}}^{\nu_U}[m_c, k_c - 1] + P_{N,\text{elec}}^{\nu_U}[m_c, k_c]} , \tag{7.11}$$

where $P_{I,\text{elec}}^{\nu_U}[m_c, k_c - 1]$ is the interference observed on the chunk $(m_c, k_c - 1)$. To ensure that the collisions are mitigated, the chunk can be assigned to user ν_U only if the *a priori* SINR estimate meets the minimum SINR target requirement, Γ_{min}, as follows:

$$\hat{\gamma}_{\nu_U}[m_c, k_c] \geq \Gamma_{\text{min}} . \tag{7.12}$$

This information can be transmitted to the serving AP either using piggyback signaling or via a dedicated control channel.

7.3.5   User scheduling and fair reservation mechanism

The score-based scheduler with reservation [129] is used for distributing N_C chunks among U_A users served by an AP. The users compete to be scheduled on chunks that are not yet reserved for any user served by a given AP. Numerical scores are calculated for each user on each chunk, which serves as a quantitative basis for ranking each user's suitability to be scheduled on each chunk. The user that is scheduled on chunk (m_c, k_c) is defined as follows:

$$\zeta_{\beta_A}[m_c, k_c] = \begin{cases} \arg\min_{v_U} s_{v_U, \beta_A}[m_c, k_c], & b_{v_U}[m_c, k_c - 1] = 0 \ \forall \ v_U \in \{1 \dots U_A\}, \\ v_U, & b_{v_U}[m_c, k_c - 1] = 1. \end{cases}$$

(7.13)

The reservation indicator on chunk (m_c, k_c), $b_{v_U}[m_c, k_c]$, is set depending on the SINR achieved at the receiver during the data transmission slot as follows:

$$b_{v_U}[m_c, k_c] = \begin{cases} 1, & \gamma_{v_U}[m_c, k_c] \geq \Gamma_{min} \ \text{and} \ \zeta_{\beta_A}[m_c, k_c] = v_U, \\ 0, & \text{otherwise}, \end{cases}$$

(7.14)

where $b_{v_U}[m_c, k_c] = 1$ indicates that the user v_U has reserved the $(m_c, k_c + 1)$ chunk by transmission of BB during the BB mini slot associated with chunk (m_c, k_c). For these reserved chunks, the modulation format is updated using the link adaptation algorithm proposed in Section 7.3.6. All users served by the tagged AP compete to get access to the idle chunks in the system. The score for each user, s_{v_U, β_A}, is calculated for each chunk by taking instantaneous fairness in the system and by respecting any restriction imposed on accessing the chunk. The score to be used in (7.13) for user v_U served by AP β_A for chunk (m_c, k_c) is calculated as follows:

$$s_{v_U, \beta_A}[m_c, k_c] = \epsilon_{v_U, \beta_A}[m_c, k_c] + \Psi_{v_U},$$

(7.15)

where the term $\epsilon_{v_U, \beta_A} \in \{0, \infty\}$ indicates whether an idle chunk (m_c, k_c) (i.e. $b_{v_U}[m_c, k_c - 1] = 0$ in (7.13)) may be accessed by user v_U, and it is given as follows:

$$\epsilon_{v_U, \beta_A}[m_c, k_c + 1] \rightarrow \begin{cases} 0, & P^{\beta_A}_{bb,elec}[m_c, k_c] \leq P_{I,elec,th} \quad \text{and} \\ & \mod(\beta_A, S) = \mod(m_c + k_c + 1, S) \\ & \text{and} \quad \hat{\gamma}_{v_U}[m_c, k_c] \geq \Gamma_{min}, \\ \infty, & \text{otherwise}. \end{cases}$$

(7.16)

In (7.16), the leftmost relation evaluates the threshold test (7.9) for interference-aware chunk selection. The middle relation determines if the AP β_A is granted access to idle chunk $(m_c, k_c + 1)$ according to the CESAR principle described in Section 7.3.4. The rightmost relation ascertains that the estimated SINR at the receiver exceeds the minimum SINR target to avoid interference from pre-established links to the tagged link. If $\epsilon_{v_U, \beta_A}[m_c, k_c + 1] \rightarrow \infty$ is set for all U_A users, the chunk remains idle in AP β_A, so that $\zeta_{\beta_A}[m_c, k_c + 1] = \emptyset$ in (7.13). In addition, the term Ψ_{v_U} is a priority penalty factor for user v_U. To ensure that the users that have fewer chunks already reserved have higher priority in accessing idle chunks than the users that have a larger number of chunks already reserved, Ψ_{ν_U} is initialized at the start of the scheduling process as follows:

$$\Psi_{\nu_U} = \exp\left(\sum_{m_c=1}^{N_C} b_{\nu_U}[m_c, k_c]\right). \tag{7.17}$$

When an idle chunk is assigned to user ν_U by (7.13), Ψ_{ν_U} is updated as follows:

$$\Psi_{\nu_U} \leftarrow \Psi_{\nu_U} \exp(1), \tag{7.18}$$

so that the users that have fewer chunks already assigned are preferred for scheduling on subsequent slots.

In the scheduler considered above, the chunks are reserved by a user by transmitting a BB as long as it has data to transmit. If the user that has successfully accessed the chunk has no more data to transmit, the reservation is annulled and the chunk may be reassigned to another user. Unfortunately, the scheduler does not preempt any user with heavy traffic volume from reserving all the chunks within the system bandwidth. Thus, such users will cause a new user entering the network or a user switching from idle state (empty transmit buffer) to active state (at least one packet in the transmit buffer) to find that all the chunks in the network are busy, thereby causing high outage. To address such a problem, a fair reservation mechanism is proposed, where the reservation made by the user by transmitting a BB is annulled by the serving AP immediately after finding that the number of chunks reserved by a user exceeds the reservation threshold. This is achieved by setting the reservation indicator $b_{\nu_U}[m_c, k_c]$ as follows:

$$b_{\nu_U}[m_c, k_c] = \begin{cases} 1, & \gamma_{\nu_U}[m_c, k_c] \geq \Gamma_{\min} \text{ and } \zeta_{\beta_A}[m_c, k_c] = \nu_U \\ & \text{and } \left(\sum_{\ell=1}^{R_{th}} b_{\nu_U}[m_c, \ell]\right) < R_{th}, \\ 0, & \text{otherwise}, \end{cases} \tag{7.19}$$

where R_{th} is the reservation threshold, which is a user-specific parameter. It is proposed that the R_{th} parameter should be set as follows:

$$R_{th} = \left\lfloor \frac{N_C}{\#\mathcal{A}_{\nu_U}[k_c]} \right\rfloor, \tag{7.20}$$

where $\lfloor \cdot \rfloor$ rounds down the number to the nearest integer smaller than the number itself, i.e. the floor operator, $\mathcal{A}_{\nu_U}[k_c]$ is the set of chunks assigned to user ν_U at time instant k_c, and $\#\mathcal{A}_{\nu_U}$ expresses the number of chunks assigned to user ν_U.

7.3.6 Link adaptation

Let $\mathcal{M} = \{1, \ldots, M_{SE}\}$ be the set of supported modulation and coding schemes. Associated with each modulation and coding scheme, $M_{SE} \in \mathcal{M}$, is a spectral efficiency, SE_M, and an SINR target $\Gamma_{M_{SE}}$ that must be achieved to satisfy a given QoS, e.g. a BER. The objective is to select the modulation scheme $M_{\nu_U}[m_c, k_c] \in \mathcal{M}$ for a chunk (m_c, k_c) that yields the highest spectral efficiency for which the condition $\gamma_{\nu_U}[m_c, k_c] \geq \Gamma_{m_{\nu_U}[m_c, k_c]}$ holds. Assuming that the channel does not change significantly between two consecutive time frames, feedback of the SINR observed in the preceding frame is used to select the appropriate modulation format for the next frame. It is sufficient to provide quantized feedback for the index of the modulation scheme using $\lceil \log_2(\#\mathcal{M}) \rceil$ bits, where $\lceil \cdot \rceil$ is the ceiling operator, and $\#\mathcal{M}$ is the cardinality of $\mathcal{M}$. Such quantized feedback may be piggybacked to the BB signal. Therefore, the additional overhead involved in dedicating one OFDM symbol to accommodate the BB mini slot is compensated by utilizing the BB for signaling purposes. In order to determine a suitable modulation format, the following steps are carried out at the UE (except step 3 which is carried out at the AP):

1. Conduct an *a priori* estimate of the SINR on chunk (m_c, k_c) based on the interference measured in the preceding frame using (7.11).
2. Determine the largest order modulation scheme, $\hat{M}_{SE}$, which fulfills the requirement $\hat{\gamma}_{v_U}[m_c, k_c] \geq \Gamma_{\hat{M}_{SE}}$ using a look-up table and feeds back the index of $\hat{M}_{SE}$ to the AP.
3. Transmit data using $M_{v_U}[m_c, k_c] = \hat{M}_{SE}$ to the intended UE.
4. Make an *a posteriori* estimate of the achieved SINR, $\gamma_{v_U}[m_c, k_c]$, using (7.5).
5. Recalculate $\hat{M}_{SE}$ using the look-up table such that $\gamma_{v_U}[m_c, k_c] \geq \Gamma_{\hat{M}_{SE}}$ holds.
6. Compute the modulation scheme to be used in the next frame as follows:

$$M_{v_U}[m_c, k_c + 1] = \begin{cases} \bar{M}_{SE}, & \gamma_{v_U}[m_c, k_c] \geq \Gamma_{M_{v_U}[m_c, k_c]+1}, \\ 0, & \gamma_{v_U} < \Gamma_{\min}, \\ M_{v_U}[m_c, k_c], & \text{otherwise}, \end{cases} \qquad (7.21)$$

where $\bar{M}_{SE} = \lceil (M_{v_U}[m_c, k_c] + \hat{M}_{SE})/2 \rceil$.

7. If $M_{v_U}[m_c, k_c + 1] = 0$, or the chunk is no longer needed, do not transmit a BB and release chunk (m_c, k_c). Otherwise, transmit a BB with the index of $M_{v_U}[m_c, k_c + 1]$ piggybacked and go to step 3.

7.3.7   System throughput with busy burst signaling

The performance of the dynamic chunk allocation mechanism using BB signaling is compared against the benchmark system using static resource partitioning according to the cluster-based approach [16, 63], similar to the one discussed in Section 7.2. However, in this section frequency reuse is considered. Here, K_C adjacent optical APs form a coordination cluster as shown in Fig. 7.5(a), and the available bandwidth is divided equally among the K_C APs within the coordination cluster as shown in Fig. 7.5(b). In this study, a cluster size $K_C = 3$ is chosen. In addition, comparisons are also made against a full chunk reuse system, where each AP transmits on the entire available bandwidth without interference coordination among neighboring cells. The performance metrics considered are the user throughput and the system throughput. The user throughput is defined as the number of bits that are received with an SINR at least equal to the SINR target required for correctly decoding the modulation format used for transmission. The system throughput is the aggregate throughput of all users served by a given AP.

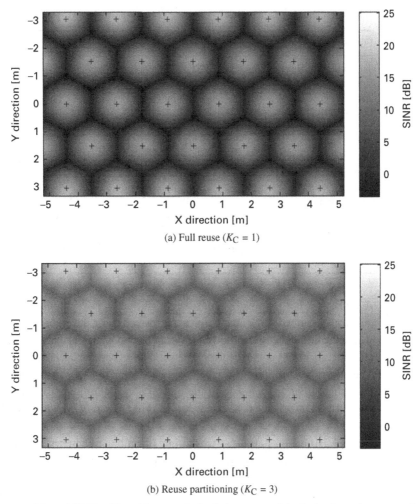

Figure 7.6 Map of SINR achieved in an optical wireless network in the horizontal cross-section of an aircraft cabin at a height of 1 m from the transmitter

In this chapter, an optical wireless network deployed inside an aircraft cabin is considered, where the APs (denoted "+") are distributed as illustrated in Fig. 7.6. The APs and the UEs employ omnidirectional transmitters and receivers. The transmitter is an array of LEDs [214], and the receiver contains an array of PDs [215]. The LEDs and the PDs have peak spectral response at 850 nm. An LED in the transmitter is directionally aligned with a PD at the receiver such that there is one active transmitter and receiver in a given direction. The components of the transmitter and the receiver are assumed to be identical both at the APs and the UEs. Furthermore, the users are uniformly distributed in a plane surface at a height of 1 m from the plane containing the APs. The optical channel path gains are modeled according to (6.3). The optical wireless channel is assumed to be frequency flat and time-invariant, which is a reasonable assumption because the dimensions of the receiver (i.e. the PD) are much larger than the carrier wavelength.

Table 7.4 Simulation parameters

FOV semi-angle of LED and PD	60°
Transmit optical power, P_{opt}	1 W
Responsivity of PD, S_{PD}	0.63 [A/W]
Area of PD, A	150 mm^2
Gain of TIA, G_{TIA}	6 kΩ
System bandwidth, B	20 MHz
Temperature, T	300 K
Background radiation, P_{bg}	6 mW/cm^2/nm
Bandwidth of optical filter	300 nm
Coding	3/4-rate convolutional
Modulation format	BPSK, QPSK, cross 8-QAM, 16-QAM, cross 32-QAM, 64-QAM, cross 128-QAM, and 256-QAM
SINR target, Γ, at 10^{-7} BER [dB]	2.2, 5.2, 9.1, 11.3, 14.4, 16.6, 19.5, 22.5

A full-buffer traffic model is considered, and perfect synchronization in time and frequency is assumed. Link adaptation is performed using the algorithm proposed in [216], where the available modulation and coding format is matched according to the prevalent channel conditions at the receiver. The simulation parameters are presented in Table 7.4. The electrical noise power is calculated according to (2.1), where the transmittance of the optical filter and the gain of the optical concentrator are assumed to be unity. The OFDMA-TDD scheme is realized by means of DCO-OFDM with $N = 2048$ subcarriers of which 2 are set to zero and the other 30 are placed within the OFDM frame for channel estimation purposes, preserving Hermitian symmetry. Therefore, $G_B = 1$ is considered. A number of subcarriers per chunk of $n_{sc} = 12$ is assumed, which yields 168 chunks per cell for full reuse and 56 chunks per cell for frequency reuse of 3. The same biasing setup as in Section 7.2.1 is employed. A positive infinite linear dynamic range is assumed, and a DC bias of 3σ is applied, resulting in negligible bottom-level clipping distortion. Therefore, $K_{\alpha_A} = 1$ and $\sigma^2_{clip,\alpha_A} = 0$ can be assumed in (7.5). As a result, $F_{O,\alpha_A} = 1$ and $F_{OE}^{\alpha_A} = F_{OE,\beta_A} = 9/10$. In addition, the DC-bias penalty is excluded from the calculation of the electrical SNR, i.e. $G_{DC}^{\alpha_A} = 1$ is considered.

The spatial distribution of the SINR around each AP is depicted for the downlink mode in Fig. 7.6. The results show that the SINR falls as low as -3.2 dB on average, given that all the available chunks are reused in each cell. However, when the allocation of chunks is coordinated using a static cluster-based resource partitioning approach, i.e. frequency reuse with $K_C = 3$, the SINR at the aforesaid location improves to approximately 8 dB. Clearly, the system is limited by interference rather than by noise for the set of system parameters considered. Therefore, the objective is to balance the reuse of chunks with the SINR achieved at the receiver such that a desirable compromise can be made between the conflicting goals of improving the system throughput and enhancing throughput at the cell edge.

In the following, the performance of BB signaling is compared against a cluster-based resource partitioning approach, as well as against a full chunk reuse approach.

For the results presented in this section, the chunk reservation policy is that a user that has transmitted a BB is assigned a chunk, as long as it has additional data to transmit in its buffer. The performance of the BB protocol is compared against the benchmark system in Fig. 7.7. The impact of the threshold parameter on the system performance is investigated using four representative values of the BB specific threshold parameter, namely $-18\,\text{dBm}$, $-15\,\text{dBm}$, $-12\,\text{dBm}$, and $-9\,\text{dBm}$. The results show that setting the threshold to $-18\,\text{dBm}$ enforces the largest exclusion region around the active receiver, and therefore leads to the highest guaranteed user throughput (measured at the 10th percentile of user throughput) among the considered thresholds and the benchmark systems. By setting the threshold to $-18\,\text{dBm}$, a guaranteed user throughput of 1.97 Mbps is achieved as shown in Fig. 7.7(b). For comparison, the static frequency planning using a reuse factor of 3 also achieves approximately the same cell-edge user throughput. This observation demonstrates that a reuse factor of 3 is optimum for the considered system at the cell edge, since the throughput cannot be improved further by reducing the reuse of the chunks. Additionally, the interference awareness property of the BB signaling mechanism helps the tagged AP to identify the chunks that are allocated to the cell-center users in the neighboring cells and to reuse such chunks to serve its own user. Therefore, the BB protocol is able to improve the throughput of cell-center users (compared to the resource partitioning approach using a reuse factor of 3) without compromising the cell-edge user throughput. As a result, the system throughput shown in Fig. 7.7(a) improves by 17% using the BB protocol compared to using the cluster-based resource partitioning approach with a cluster size of 3.

When the threshold is gradually increased, the system throughput increases due to an increase in the spatial reuse of chunks, but the throughput at the cell edge decreases. Initially, the increase in CCI reduces the achieved SINR at the receiver and forces the transmitters to utilize a lower order modulation and coding format. However, when further increasing the threshold, the cell-edge users are forced to release the reserved chunks, since the minimum SINR target is no longer met. The scheduler then assigns the released chunks to cell-center users, where the minimum SINR target is more likely to be met. This is because the CCI caused to the receiver from a pre-established transmission in the neighboring cell causes the minimum SINR target from (7.11) not to be met for the cell-edge users. As a result, the number of chunks allocated to users close to the serving AP increase but the number of chunks available to the cell-edge users is reduced. Hence, the system throughput increases as shown in Fig. 7.7(a) at the cost of the cell-edge user throughput shown in Fig. 7.7(b). In particular, when the interference protection is annulled by setting high thresholds, such as $-9\,\text{dBm}$, the performance at the cell edge approaches that of a full reuse system as shown in Fig. 7.7(b). The released chunks are reallocated to cell-center users, which improves the system throughput but with outage at the cell edge. The results in Fig. 7.7 depict the various degrees of trade-off between interference protection and spatial reuse simply by adjusting the threshold parameter.

The results in Fig. 7.7 are obtained assuming full-buffer traffic, which represents an upper limit on the offered load, where each user has enough data to fill the entire bandwidth in every slot. For a practical application, such as accessing multimedia contents, the user traffic is composed of Internet protocol (IP) packets. An interesting question

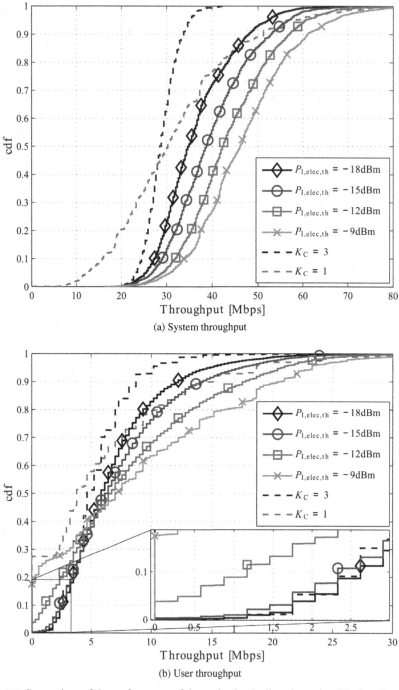

(a) System throughput

(b) User throughput

Figure 7.7 Comparison of the performance of dynamic chunk allocation using BB signaling in an optical wireless network

would be to quantify the performance of the proposed approach, when the traffic is bursty. To investigate the performance, the system is simulated for various offered loads, ranging from approximately 10 Mbps/cell to 160 Mbps/cell, and the results are presented in Figs. 7.8 and 7.9. The offered loads are varied by adjusting the inter-arrival time, which follows an exponential distribution. The results show that interference coordination, whether that is done statically or dynamically, improves the performance of the system compared to that achieved without interference coordination, i.e. full chunk reuse. Moreover, it can also be observed that dynamic chunk allocation using the BB protocol achieves better performance as compared to what is achieved with the cluster-based static chunk allocation using a cluster size of 3 in terms of median user throughput, median system throughput, and peak user throughput, as shown in Figs. 7.8 and 7.9(a). This is attributed to the interference awareness property of the BB protocol that allows the system to identify the chunks that can be reused in the tagged cell without causing detrimental CCI to the users served by the neighboring APs. However, an interesting behavior can be observed with the guaranteed user throughput, i.e. the 10th percentile of user throughput in Fig. 7.9(b). When offered loads less than 25 Mbps/cell are imposed on the system, the chunks in the system are not fully reused, and therefore chunk allocation using the BB approach provides the same throughput as the static chunk allocation approach. When the offered load is increased from 25 Mbps/cell up to 40 Mbps/cell, dynamic chunk allocation using the BB protocol delivers a superior performance in terms of median user throughput, median system throughput, and peak user throughput as shown in Figs. 7.8 and 7.9(a), whilst achieving the same guaranteed user throughput shown in Fig. 7.9(b). Furthermore, the interference threshold can be adjusted dynamically to trade off between the aggregate system throughput and through-put guaranteed at the cell edge.

However, when offered loads higher than 40 Mbps are imposed on the system, the guaranteed user throughput with the BB protocol degrades with increases in the offered load. By contrast, the guaranteed user throughput with static chunk allocation increases until a peak is reached, and generally the same level of throughput is maintained. Note that the considered system bandwidth is only 20 MHz, out of which only 10 MHz can be allocated independently. The remaining 10 MHz bandwidth is used to transmit the complex conjugate of the data symbols in order to maintain a real-valued signal. With these parameters, the peak raw data rate in an isolated cell would be 10 Mbps, assuming a binary phase shift keying (BPSK) modulation and 80 Mbps assuming 256-QAM. Therefore, this reflects a scenario where a user with heavy traffic demand competes with other users that may possibly have heavy or light traffic demands. Provided that the user with heavy traffic demand has successfully accessed the chunk and reserved it by transmitting a BB signal, the chunks available at the AP are exclusively assigned to the user with the heavy demand. Such chunks appear unavailable to a user that has just entered the network or switched from idle (empty transmit buffer) to active (containing at least one protocol data unit (PDU) queued in the transmit buffer). Likewise, an active user that releases a chunk, when its transmit buffer is empty, will find that the chunks are all occupied at a later point in time when the user attempts to transmit data again. Moreover, increasing the traffic load leads to a higher number of frames, where a chunk reserved by a user appears unavailable to other users. Thus, the ability of a user to

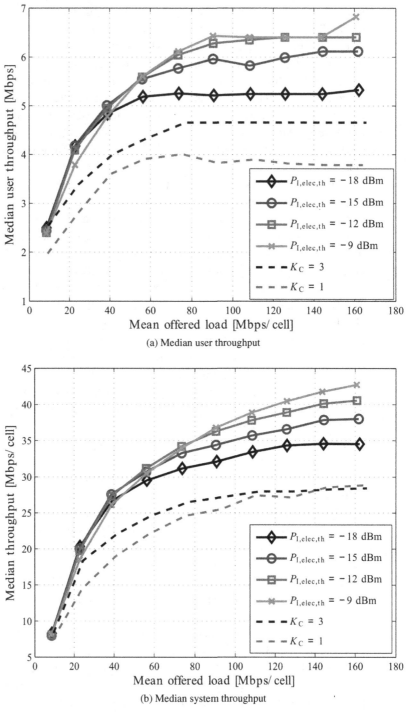

(a) Median user throughput

(b) Median system throughput

Figure 7.8 Median user throughput and median system throughput vs. mean offered load of chunk allocation with BB signaling and static resource partitioning, where the successfully accessed chunks are reserved until the transmit buffer is empty

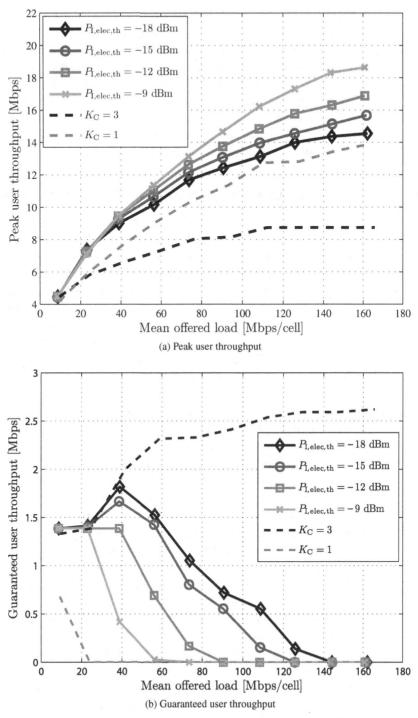

(a) Peak user throughput

(b) Guaranteed user throughput

Figure 7.9 Peak user throughput and guaranteed user throughput vs. mean offered load of chunk allocation with BB signaling and static resource partitioning, where the successfully accessed chunks are reserved until the transmit buffer is empty

re-acquire the released chunks decreases with increasing the traffic load. Therefore, the guaranteed user throughput decreases with increasing the offered load.

The assumption that the reserved chunk will be allocated to the same user in the next slot ensures that the *a priori* knowledge of the amount of CCI caused to the user served by a neighboring cell is valid. However, the results presented in Figs. 7.8 and 7.9 have demonstrated that allowing for the reservation of chunks until the transmit buffer is emptied deteriorates system performance, once the system is overloaded. To address this shortcoming, the fair chunk reservation mechanism proposed in Section 7.3.5 is employed, and its performance is analyzed in the next section.

7.3.8   System throughput with busy burst signaling and fair reservation mechanism

The performance of the BB protocol with the fair reservation policy is presented in Figs. 7.10 and 7.11. By allowing each user to reserve the successfully accessed chunks for a duration of time inversely proportional to the amount of bandwidth occupied, the problem of outage due to unavailability of idle chunks is avoided. Consequently, the guaranteed user throughput shown in Fig. 7.11(b) does not deteriorate, when the offered load is increased, as long as the threshold parameter is set so as to enable the cell-edge users to meet their minimum SINR targets. Interestingly, when the average traffic load in the system is less than 40 Mbps/cell, the performance of dynamic chunk allocation using the BB protocol is the same regardless of whether or not the fair scheduling is applied. This is because at such offered loads, the user buffer is emptied periodically, and the users release the chunks they have reserved. Therefore, the chunks are still available to users that enter the network at a later point in time.

When the offered load in the system is increased to 40 Mbps/cell, the BB protocol with fair reservation achieves up to 14% higher guaranteed user throughput, shown in Fig. 7.11(b), together with approximately 13% increase in median system throughput, shown in Fig. 7.10(b), as compared to what is achieved with the static chunk allocation algorithm. Moreover, it can be observed that the dynamic chunk allocation with BB continues to outperform the static resource partitioning approach both in terms of guaranteed user throughput and median system throughput until an offered load of 100 Mbps/cell is imposed on the system. This demonstrates that the BB protocol with a fair reservation policy allows the spectrum to be shared more flexibly among competing users and across the APs in the network, thereby improving the spectral efficiency. By contrast, the static resource allocation leaves some of the chunks unoccupied in the APs, where the instantaneous offered load is low, while it is unable to cope with the high instantaneous offered load in other APs. It should be noted that the difference between the guaranteed user throughput obtained using the BB protocol with fair reservation and the one obtained using resource partitioning decreases as the offered load increases, and eventually the static resource partitioning approach slightly exceeds the performance of the former. This is attributed to the fact that the number of chunks that remain idle decreases with an increase in the offered load, eventually leading to no idle chunks within the bandwidth assigned to the tagged AP, when the static chunk allocation approach with cluster-based resource partitioning is used. By contrast, with

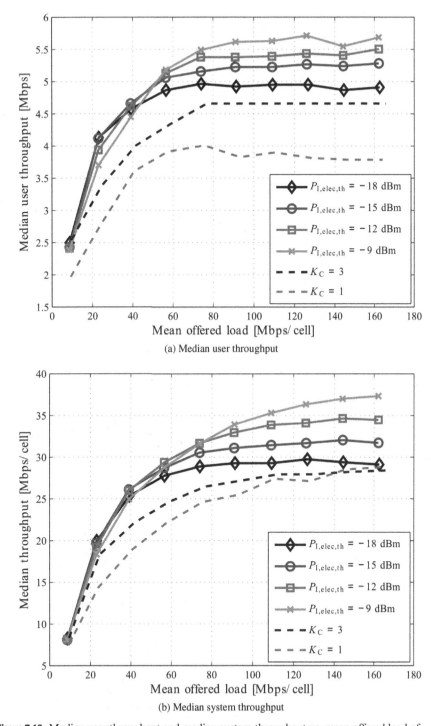

(a) Median user throughput

(b) Median system throughput

Figure 7.10 Median user throughput and median system throughput vs. mean offered load of chunk allocation with BB signaling and static resource partitioning, where the successfully accessed chunks are reserved for a period of time inversely proportional to the amount of bandwidth available to the user

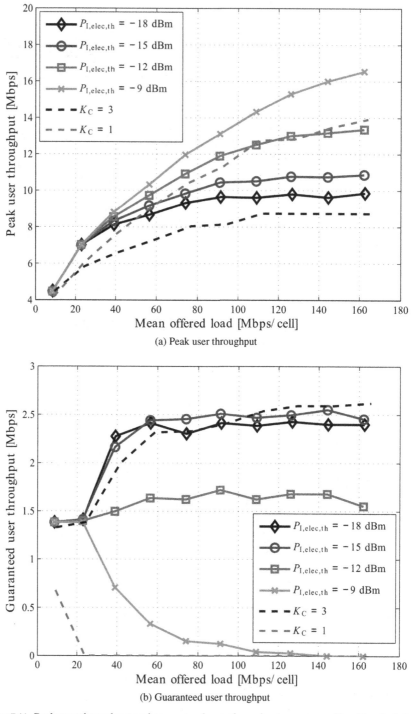

(a) Peak user throughput

(b) Guaranteed user throughput

Figure 7.11 Peak user throughput and guaranteed user throughput vs. mean offered load of chunk allocation with BB signaling and static resource partitioning, where the successfully accessed chunks are reserved for a period of time inversely proportional to the amount of bandwidth available to the user

the BB signaling approach, some of the chunks remain idle for up to three frames, when they are released in order to conform to the fair reservation policy. Furthermore, the results obtained using the full-buffer traffic model have established that a reuse factor of 3 is an ideal reuse factor at the cell edge as shown in Fig. 7.7(b). Due to the above two facts, the performance of the BB protocol with fair reservation degrades slightly when the offered load gradually approaches the upper bound as shown Fig. 7.11(b). Likewise, the median system throughput obtained with the BB protocol degrades as shown in Figs. 7.8(b) and 7.10(b) when the fair reservation policy is applied, given that the same threshold is used both with and without the fair reservation policy. This is partly attributed to the fact that the chunk is released after reaching the reservation threshold and partly to the fact that some chunks must remain unused in the tagged cell to avoid causing detrimental CCI to the cell-edge users served in the neighboring cell.

In summary, the results have established that the BB protocol with fair reservation policy provides a scalable mechanism to flexibly share the available chunks in the system by adjusting the threshold parameter to reach a desired balance between the conflicting goals of enhancing spectral efficiency against improving quality of service in the system. This self-organizing property is particularly important for deployment scenarios such as the data access networks in an aircraft cabin, where the density of users is particularly high, and the network potentially needs to cater for a wide variety of traffic classes.

7.4   Summary

In this chapter, first, the throughput of a cellular OFDM-based OWC system with static resource partitioning according to wavelength reuse has been simulated inside the cabin of an aircraft by means of a global MCRT irradiation simulation using a CAD cabin model and assuming off-the-shelf front-end components. After O/E conversion of the spatial irradiance distribution, electrical SINR maps have been obtained for wavelength reuse factors of 1, 2, and 3. In addition, an AMC scheme was applied to the two considered O-OFDM system realizations, i.e. ACO-OFDM and DCO-OFDM, in order to translate the SINR maps into transmission rates at a given target BER. The ACO-OFDM system supported higher data rates at the cell edge, while DCO-OFDM doubled the throughput towards the cell center, manifesting the trade-off between power efficiency and spectral efficiency. In the presented network configuration, both transmission schemes required wavelength reuse factors of 3 or higher in order to ensure full cell support.

In addition, dynamic interference-aware self-organizing resource allocation using the BB protocol has been considered in a cellular OWC network with OFDMA-TDD deployed inside an aircraft cabin. The performance of the proposed BB approach was compared against that of a chunk allocation approach, where interference was avoided through the use of cluster-based static resource partitioning. Compared to static resource partitioning using a frequency reuse factor of 3, the BB approach improved the median system throughput by 17%, while maintaining an equivalent throughput at the cell edge, when assuming a full-buffer traffic model. Moreover, it was demonstrated that the BB specific threshold parameter could be adjusted to trade off cell-edge user throughput for aggregate system throughput and vice versa. In this context, either a median system throughput of 46 Mbps/cell or a cell-edge user throughput of 2.6 Mbps was demonstrated to be feasible with the BB protocol using a total system bandwidth of 20 MHz. A heuristic for fair reservation of chunks was proposed that mitigates the problem of outage arising when the users with heavy traffic demands reserve the available chunks, leaving no chunks available to the new users who try to access the network. The results demonstrated that the proposed method outperformed the static cluster-based resource partitioning approach approximately by 14% both in terms of median system throughput and guaranteed user throughput. Since the decision whether or not to reuse a chunk was based on comparing the received BB power against a threshold value, the network could be dynamically reconfigured to satisfy the guaranteed data rates or BER of different service classes simply by adjusting the threshold parameter. This self-organizing property was deemed particularly important in an optical wireless network deployed inside an aircraft cabin, which was characterized by high user density and heterogeneous traffic demands.

References

[1] The Flag Press, "History Behind Semaphore Flags," retrieved from http://flagexpressions.
wordpress.com/2010/03/23/history-behind-semaphore-flags/, Nov. 17, 2012.

[2] A. G. Bell, "Selenium and the Photophone," *Nature*, vol. 22, no. 569, pp. 500–5030,
1880.

[3] F. R. Gfeller and U. Bapst, "Wireless In-House Data Communication Via Diffuse Infrared
Radiation," *Proceedings of the IEEE*, vol. 67, no. 11, pp. 1474–1486, Nov. 1979.

[4] Z. Ghassemlooy, W. Popoola, and S. Rajbhandari, *Optical Wireless Communications:
System and Channel Modeling with MATLAB(R)*, 1st edn. CRC Press, 2012.

[5] G. W. Marsh and J. M. Kahn, "Performance Evaluation of Experimental 50-Mb/s Diffuse
Infrared Wireless Link Using On–Off Keying with Decision-Feedback Equalization,"
IEEE Transactions on Communications, vol. 44, no. 11, pp. 1496–1504, Nov. 1996.

[6] J. B. Carruthers and J. M. Kahn, "Angle Diversity for Nondirected Wireless Infrared
Communication," *IEEE Transactions on Communications*, vol. 48, no. 6, pp. 960–969,
June 2000.

[7] Y. Tanaka, T. Komine, S. Haruyama, and M. Nakagawa, "Indoor Visible Light Data Transmission System Utilizing White LED Lights," *IEICE Transactions on Communications*,
vol. E86-B, no. 8, pp. 2440–2454, Aug. 2003.

[8] M. Afgani, H. Haas, H. Elgala, and D. Knipp, "Visible Light Communication Using
OFDM," in *Proceedings of the 2nd International Conference on Testbeds and Research
Infrastructures for the Development of Networks and Communities (TRIDENTCOM)*,
Barcelona, Spain, March 1–3, 2006, pp. 129–134.

[9] J. Vucic, C. Kottke, S. Nerreter, K. D. Langer, and J. W. Walewski, "513 Mbit/s Visible
Light Communications Link Based on DMT-Modulation of a White LED," *Journal of
Lightwave Technology*, vol. 28, no. 24, pp. 3512–3518, Dec. 2010.

[10] J. Vucic, C. Kottke, K. Habel, and K. Langer, "803 Mbit/s Visible Light WDM Link Based
on DMT Modulation of a Single RGB LED Luminary," in *Optical Fiber Communication
Conference and Exposition (OFC/NFOEC)*, March 2011, pp. 1–3.

[11] A. M. Khalid, G. Cossu, R. Corsini, P. Choudhury, and E. Ciaramella, "1-Gb/s Transmission Over a Phosphorescent White LED by Using Rate-Adaptive Discrete Multitone
Modulation," *IEEE Photonics Journal*, vol. 4, no. 5, pp. 1465–1473, Oct. 2012.

[12] G. Cossu, A. M. Khalid, P. Choudhury, R. Corsini, and E. Ciaramella, "3.4 Gbit/s Visible
Optical Wireless Transmission Based on RGB LED," *Optics Express*, vol. 20, pp. B501–
B506, 2012.

[13] A. Azhar, T. Tran, and D. O'Brien, "A Gigabit/s Indoor Wireless Transmission Using
MIMO-OFDM Visible-Light Communications," *IEEE Photonics Technology Letters*,
vol. 25, no. 2, pp. 171–174, Jan. 15, 2013.

[14] D. Tsonev, H. Chun, S. Rajbhandari, J. McKendry, S. Videv, E. Gu, M. Haji, S. Watson, A. Kelly, G. Faulkner, M. Dawson, H. Haas, and D. O'Brien, "A 3-Gb/s Single-LED OFDM-based Wireless VLC Link Using a Gallium Nitride μLED," *IEEE Photonics Technology Letters*, vol. 99 (to appear), p. 4, 2014.

[15] "'Li-fi' via LED Light Bulb Data Speed Breakthrough," BBC News, Oct. 2013, http://www.bbc.co.uk/news/technology-24711935.

[16] S. Dimitrov, R. Mesleh, H. Haas, M. Cappitelli, M. Olbert, and E. Bassow, "On the SIR of a Cellular Infrared Optical Wireless System for an Aircraft," *IEEE Journal on Selected Areas in Communications (IEEE JSAC)*, vol. 27, no. 9, pp. 1623–1638, Dec. 2009.

[17] S. Dimitrov, H. Haas, M. Cappitelli, and M. Olbert, "On the Throughput of an OFDM-based Cellular Optical Wireless System for an Aircraft Cabin," in *Proceedings of the European Conference on Antennas and Propagation (EuCAP 2011)*, Rome, Italy, April 11–15, 2011, invited Paper.

[18] B. Ghimire and H. Haas, "Self Organising Interference Coordination in Optical Wireless Networks," *EURASIP Journal on Wireless Communications and Networking*, vol. 1, no. 131, April 2012.

[19] H. Haas, "Wireless Data from Every Light Bulb," TED Website, Aug. 2011. http://bit.ly/tedvlc.

[20] J. R. Barry, *Wireless Infrared Communications*. Springer, 1994, vol. 280.

[21] J. M. Kahn and J. R. Barry, "Wireless Infrared Communications," *Proceedings of the IEEE*, vol. 85, no. 2, pp. 265–298, Feb. 1997.

[22] C. Singh, J. John, Y. N. Singh, and K. K. Tripathi, "A Review of Indoor Optical Wireless Systems," *IETE Technical Review*, vol. 19, pp. 3–17, Jan.–April 2002.

[23] P. Barker and A. C. Boucouvalas, "Performance Modeling of the IrDA Protocol for Infrared Wireless Communications," *IEEE Communication Magazine*, vol. 36, no. 12, pp. 113–117, 1998.

[24] European Commission, "Technical Briefing: Phasing out Incandescent Bulbs in the EU," retrieved from http://ec.europa.eu, Sept. 2008.

[25] T. Komine and M. Nakagawa, "Fundamental Analysis for Visible-Light Communication System using LED Lights," *IEEE Transactions on Consumer Electronics*, vol. 50, no. 1, pp. 100–107, Feb. 2004.

[26] —— "Performance Evaluation of Visible-Light Wireless Communication System using White LED Lightings," in *Proceedings of the ISCC 2004. Ninth-International Symposium on Computers and Communications*, vol. 1, June 28– July 1, 2004, pp. 258– 263.

[27] M. Miki, E. Asayama, and T. Hiroyasu, "Visible-Light Communication using Visible-Light Communication Technology," in *Proceedings of the IEEE Conference on Cybernetics and Intelligence Systems (CIS 06)*, Bangkok, Thailand, June 7–9, 2006, pp. 1–6.

[28] J. Grubor, S. Randel, K. Langer, and J. W. Walewski, "Broadband Information Broadcasting Using LED-Based Interior Lighting," *Journal of Lightwave Technology*, vol. 26, pp. 3883–3892, 2008.

[29] D. C. O'Brien, L. Zeng, H. Le-Minh, G. Faulkner, J. W. Walewski, and S. Randel, "Visible Light Communications: Challenges and Possibilities," in *IEEE International Symposium on Personal, Indoor and Mobile Radio Communications*, ser., Cannes, France, Sept. 2008.

[30] H. Elgala, R. Mesleh, and H. Haas, "Indoor Broadcasting via White LEDs and OFDM," *IEEE Transactions on Consumer Electronics*, vol. 55, no. 3, pp. 1127–1134, Aug. 2009.

[31] —— "Indoor Optical Wireless Communication: Potential and State-of-the-Art," *IEEE Commun. Mag.*, vol. 49, no. 9, pp. 56–62, Sept. 2011.

[32] D. O'Brien, "Visible Light Communications: Challenges and Potential," in *Proceedings of the IEEE Photonics Conference (PHO)*, Virginia, USA, Oct. 2011, pp. 365–366.

[33] L. Hanzo, H. Haas, S. Imre, D. O'Brien, M. Rupp, and L. Gyongyosi, "Wireless Myths, Realities and Futures: From 3G/4G to Optical and Quantum Wireless," *Proceedings of the IEEE*, vol. 100, pp. 1853–1888, May 2012.

[34] Visible Light Communications Consortium (VLCC), retrieved from http://www.vlcc.net, Feb. 2007.

[35] IEEE Std. 802.15.7-2011, *IEEE Standard for Local and Metropolitan Area Networks, Part 15.7: Short-Range Wireless Optical Communication Using Visible Light*, IEEE Std., 2011.

[36] J. G. Andrews, A. Ghosh, and R. Muhamed, *Fundamentals of WiMAX*. Pearson Education, 2007.

[37] BROADCOM Corporation, "802.11n: Next-Generation Wireless LAN Technology," White paper, BROADCOM Corporation, Tech. Rep., April 2006, retrieved from http://www.broadcom.com/docs/WLAN/802-11n-WP100-R.pdf, Aug. 4, 2006.

[38] Ofcom, "Study on the Future UK Spectrum Demand for Terrestrial Mobile Broadband Applications," Realwireless, report, June 2013.

[39] "Visible Light Communication (VLC)–A Potential Solution to the Global Wireless Spectrum Shortage," GBI Research, Tech. Report, 2011, http://www.gbiresearch.com/.

[40] Cisco Visual Networking Index, "Global Mobile Data Traffic Forecast Update, 2012-2017," CISCO, White Paper, Feb. 2013, http://www.cisco.com/en/US/solutions/collateral/ns341/ns525/ns537/ns705/ns827/white_paper_c11-481360.pdf.

[41] Wireless Gigabit Aliance, "Specifications," retrieved from http://wirelessgigabitalliance.org/specifications/, Nov. 19, 2012.

[42] IEEE P802.11 - Task Group AD, "Very High Throughput in 60 GHz," retrieved from http://www.ieee802.org/11/Reports/tgad_update.htm, Nov. 19, 2012.

[43] S. Dimitrov and H. Haas, "Optimum Signal Shaping in OFDM-based Optical Wireless Communication Systems," in *Proceeding of the IEEE Vehicular Technology Conference (IEEE VTC Fall)*, Quebec City, Canada, Sept. 3–6, 2012.

[44] BS EN 62471:2008, *Photobiological Safety of Lamps and Lamp Systems*, BSI British Standards Std., Sept. 2008.

[45] J. G. Proakis, *Digital Communications*, 4th edn. McGraw-Hill, 2000.

[46] H. Haas and P. Valtink, *100 Produkte der Zukunft–Wegweisende Ideen, die unser Leben verändern werden*. Econ Verlag, 2007, ch. Sprechendes Licht–Leuchtdioden übermitteln Daten.

[47] H. Elgala, R. Mesleh, H. Haas, and B. Pricope, "OFDM Visible Light Wireless Communication Based on White LEDs," in *Proceedings of the 64th IEEE Vehicular Technology Conference (VTC)*, Dublin, Ireland, April 22–25, 2007.

[48] H. Burchardt, N. Serafimovski, D. Tsonev, S. Videv, and H. Haas, "VLC: Beyond Point-to-Point Communication," *IEEE Communications Magazine*, 2014, pre-print, http://www.eng.ed.ac.uk/drupal/hxh/publications/.

[49] J. Campello, "Practical Bit Loading for DMT," in *Proceedings of the IEEE International Conference on Communications (IEEE ICC 1999)*, vol. 2, Vancouver, BC, Canada, June 6–10, 1999, pp. 801–805.

[50] H. E. Levin, "A Complete and Optimal Data Allocation Method for Practical Discrete Multitone Systems," in *Proceedings of the IEEE Global Telecommunications Conference (IEEE GLOBECOM 2001)*, vol. 1, San Antonio, TX, USA, Nov. 25–29, 2001, pp. 369–374.

[51] J. Armstrong and A. Lowery, "Power Efficient Optical OFDM," *Electronics Letters*, vol. 42, no. 6, pp. 370–372, March 16, 2006.

[52] S. C. J. Lee, S. Randel, F. Breyer, and A. M. J. Koonen, "PAM-DMT for Intensity-Modulated and Direct-Detection Optical Communication Systems," *IEEE Photonics Technology Letters*, vol. 21, no. 23, pp. 1749–1751, Dec. 2009.

[53] N. Fernando, Y. Hong, and E. Viterbo, "Flip-OFDM for Optical Wireless Communications," in *Information Theory Workshop (ITW)*. Paraty, Brazil: IEEE, Oct. 16–20, 2011, pp. 5–9.

[54] D. Tsonev, S. Sinanović, and H. Haas, "Novel Unipolar Orthogonal Frequency Division Multiplexing (U-OFDM) for Optical Wireless," in *Proceedings of the Vehicular Technology Conference (VTC Spring)*, Yokohama, Japan: IEEE, May 6–9, 2012.

[55] K. Asadzadeh, A. Farid, and S. Hranilovic, "Spectrally Factorized Optical OFDM," in *12th Canadian Workshop on Information Theory (CWIT 2011)*. IEEE, May 17–20, 2011, pp. 102–105.

[56] J. Fakidis, D. Tsonev, and H. Haas, "A Comparison between DCO-OFDMA and Synchronous One-dimensional OCDMA for Optical Wireless Communications," in *2013 IEEE 24th International Symposium on Personal Indoor and Mobile Radio Communications (PIMRC)*, London, Sept. 8–11, 2013, pp. 3605–3609.

[57] M. B. Rahaim, A. M. Vegni, and T. D. C. Little, "A Hybrid Radio Frequency and Broadcast Visible Light Communication System," in *IEEE Global Communications Conference (GLOBECOM 2011) Workshops*, Dec. 5-9, 2011, pp. 792–796.

[58] pureVLC. pureVLC Li-1st. video. http://purevlc.co.uk/li-fire/purevlc-li-1st/

[59] S. Ortiz, "The Wireless Industry Begins to Embrace Femtocells," *Computer*, vol. 41, no. 7, pp. 14–17, July 2008.

[60] V. Chandrasekhar, J. Andrews, and A. Gatherer, "Femtocell Networks: A Survey," *IEEE Communications Magazine*, vol. 46, no. 9, pp. 59–67, 2008.

[61] H. Haas, "High-Speed Wireless Networking Using Visible Light," retrieved from https://spie.org/x93593.xml, 2013.

[62] I. Stefan, H. Burchardt, and H. Haas, "Area Spectral Efficiency Performance Comparison between VLC and RF Femtocell Networks," in *Proceedings of the International Conference on Communications (ICC)*, Budapest, Hungary, Jun. 2013, pp. 1–5.

[63] G. W. Marsh and J. M. Kahn, "Channel Reuse Strategies for Indoor Infrared Wireless Communications," *IEEE Transactions on Communications*, vol. 45, no. 10, pp. 1280–1290, Oct. 1997.

[64] C. Chen, N. Serafimovski, and H. Haas, "Fractional Frequency Reuse in Optical Wireless Cellular Networks," in *Proceedings of the IEEE International Symposium on Personal, Indoor and Mobile Radio Communications (PIMRC 2013)*. London: IEEE, Sept. 8–11, 2013.

[65] C. Chen, D. Tsonev, and H. Haas, "Joint Transmission in Indoor Visible Light Communication Downlink Cellular Networks," in *Proceedings of the IEEE Workshop on Optical Wireless Communication (OWC 2013)*, Atlanta, GA: IEEE, Dec. 9, 2013.

[66] S. Dimitrov, R. Mesleh, H. Haas, M. Cappitelli, M. Olbert, and E. Bassow, "Path Loss Simulation of an Infrared Optical Wireless System for Aircraft," in *Proceedings of the IEEE Global Communications Conference (IEEE GLOBECOM 2009)*, Honolulu, HI, Nov. 30–Dec. 4, 2009.

[67] S. Dimitrov, S. Sinanovic, and H. Haas, "Signal Shaping and Modulation for Optical Wireless Communication," *IEEE/OSA Journal on Lightwave Technology (IEEE/OSA JLT)*, vol. 30, no. 9, pp. 1319–1328, May 2012.

[68] F. R. Gfeller, P. Bernasconi, W. Hirt, C. Elisii, and B. Weiss, "Dynamic Cell Planning for Wireless Infrared In-House Data Transmission," *Mobile Communications Advanced Systems and Components*, ser. Lecture Notes in Computer Science. Springer Berlin/Heidelberg, 1994, vol. 783, pp. 261–272.

[69] S. B. Alexander, *Optical Communication Receiver Design*, 1st edn. SPIE Press Book, Jan. 1997.

[70] D. C. O'Brien, G. E. Faulkner, S. Zikic, and N. P. Schmitt, "High Data-Rate Optical Wireless Communications in Passenger Aircraft: Measurements and Simulations," in *6th International Symposium on Communication Systems, Networks and Digital Signal Processing (CSNDSP'08)*, Graz, Austria, July 23–25, 2008, pp. 68–71.

[71] G. Yun and M. Kavehrad, "Indoor Infrared Wireless Communications Using Spot Diffusing and Fly-Eye Receivers," *The Canadian Journal of Electrical and Computer Engineering*, vol. 18, no. 4, Oct. 1993.

[72] M. Kavehrad and S. Jivkova, "Indoor Broadband Optical Wireless Communications: Optical Subsystems Designs and their Impact on Channel Characteristics," *IEEE Wireless Communications Magazine*, vol. 10, no. 2, pp. 30–35, 2003.

[73] F. E. Alsaadi and J. M. H. Elmirghani, "Spot Diffusing Angle Diversity MC-CDMA Optical Wireless System," in *Proceedings of the IET Optoelectronics*, vol. 3, no. 3, pp. 131–141, June 2009.

[74] T. Borogovac, M. Rahaim, and J. B. Carruthers, "Spotlighting for Visible Light Communications and Illumination," in *Proceedings of the IEEE Global Communications Conference (GLOBECOM 2010) Workshops*, Dec. 6-10, 2010, pp. 1077–1081.

[75] J. Kahn, J. Barry, W. Krause, M. Audeh, J. Carruthers, G. Marsh, E. Lee, and D. Messerschmitt, "High-Speed Non-Directional Infrared Communication for Wireless Local-Area Networks," in *Proceedings of the 26th Asilomar Conference on Signals, Systems and Computers*, vol. 1, California, Oct. 26–28, 1992, pp. 83–87.

[76] J. M. Kahn, W. J. Krause, and J. B. Carruthers, "Experimental Characterization of Non-Directed Indoor Infrared Channels," *IEEE Transactions on Communications*, vol. 43, no. 234, pp. 1613–1623, Feb.–Mar.–Apr. 1995.

[77] J. B. Carruthers and J. M. Kahn, "Modeling of Nondirected Wireless Infrared Channels," in *Proceedings of the IEEE Conference on Communications: Converging Technologies for Tomorrow's Applications*, vol. 2, Dallas, TX, June 23–27, 1996, pp. 1227–1231.

[78] M. R. Pakravan, M. Kavehrad, and H. Hashemi, "Indoor Wireless Infrared Channel Characterization by Measurements," *IEEE Transactions on Vehicular Technology*, vol. 50, no. 4, pp. 1053–1073, July 2001.

[79] D. C. O'Brien, S. H. Khoo, W. Zhang, G. E. Faulkner, and D. J. Edwards, "High-speed Optical Channel Measurement System," in *Proceedings of the SPIE, Optical Wireless Communications IV*, vol. 4530, Denver, CO, Aug. 21–22, 2001, pp. 135–144.

[80] K. Smitha and J. John, "Propagation Measurements of Indoor Infrared Channels," Dec. 2004, retrieved from http://ee.iust.ac.ir/profs/Sadr/Papers/ltwp20.pdf, Dec. 29, 2004.

[81] J. Barry, J. Kahn, W. Krause, E. Lee, and D. Messerschmitt, "Simulation of Multipath Impulse Response for Indoor Wireless Optical Channels," *IEEE Journal on Selected Areas in Communication*, vol. 11, no. 3, pp. 367–379, April 1993.

[82] A. Muller, "Monte-Carlo Multipath Simulation of Ray Tracing Channel Models," in *Proceedings of the IEEE Global Telecommunications Conference GLOBECOM 1994, Communications: The Global Bridge*, vol. 3, Nov. 28–Dec. 2, 1994, pp. 1446–1450.

[83] F. Lopez-Hernandez, R. Perez-Jimenez, and A. Santamaria, "Ray-tracing Algorithms for Fast Calculation of the Impulse Response on Diffuse IR-wireless Indoor Channels," *Optical Engineering*, vol. 39, no. 10, pp. 2775–2780, Oct. 2000.

[84] V. Pohl, V. Jungnickel, and C. von Helmolt, "A Channel Model for Wireless Infrared Communication," in *Proceedings of the 11th IEEE International Symposium on Personal, Indoor and Mobile Radio Communications PIMRC 2000*, vol. 1, London, Sept. 18–21, 2000, pp. 297–303.

[85] V. Jungnickel, V. Pohl, S. Nonnig, and C. von Helmolt, "A Physical Model of the Wireless Infrared Communication Channel," *IEEE Journal on Selected Areas in Communications*, vol. 20, no. 3, pp. 631–640, April 2002.

[86] S. Rodriguez, R. Perez-Jimenez, F. Lopez-Hernandez, O. Gonzalez, and A. Ayala, "Reflection Model for Calculation of the Impulse Response on IR-wireless Indoor Channels Using Ray-tracing Algorithm," *Microwave Optical Technology Letter*, vol. 32, no. 4, pp. 296–300, Jan. 2002.

[87] O. Gonzalez, S. Rodriguez, R. Perez-Jimenez, B. R. Mendoza, and A. Ayala, "Error Analysis of the Simulated Impulse Response on Indoor Wireless Optical Channels Using a Monte Carlo-based Ray-tracing Algorithm," *IEEE Transactions on Communications*, vol. 53, no. 1, pp. 124–130, Jan. 2005.

[88] H. Naoki and I. Takeshi, "Channel Modeling of Non-Directed Wireless Infrared Indoor Diffuse Link," *Electronics and Communications in Japan (Part I: Communications)*, vol. 90, no. 6, pp. 9–19, Feb. 2007.

[89] G. Ntogari, T. Kamalakis, and T. Sphicopoulos, "Performance Analysis of Non-directed Equalized Indoor Optical Wireless Systems," in *Proceedings of the 6th International Symposium on Communication Systems, Networks and Digital Signal Processing CSNDSP 2008*, Graz, Austria, July 23–25, 2008, pp. 156–160.

[90] M. Bertrand, O. Bouchet, and P. Besnard, "Personal Optical Wireless Communications: LOS/WLOS/DIF Propagation Model and QOFI," in *Proceedings of the 6th International Symposium on Communication Systems, Networks and Digital Signal Processing CSNDSP 2008*, Graz, Austria, July 23–25, 2008, pp. 179–182.

[91] Vishay Semiconductors, "Datasheet: TSHG8200 High Speed Infrared Emitting Diode, 830 nm, GaAlAs Double Hetero," retrieved July 26, 2011 from http://www.vishay.com/docs/84755/tshg8200.pdf, July 2008.

[92] Osram Opto Semiconductors, "Datasheet: LCW W5SM Golden Dragon White LED," retrieved from http://www.osram.de, April 2011.

[93] OSRAM GmbH, "Datasheet: OS-PCN-2008-002-A OSTAR LED," retrieved from http://www.osram.de, Feb. 2008.

[94] Vishay Semiconductors, "Infrared Emitters," retrieved from http://www.vishay.com/ir-emitting-diodes/, Jan. 25, 2011.

[95] Hamamatsu Photonics K. K., "Silicon PIN Photodiode," retrieved from http://sales.hamamatsu.com/en/products/solid-state-division/si-photodiode-series/si-pin-photodiode.php, Jan. 25, 2011.

[96] Thorlabs, "Bandpass Filters," retrieved from http://www.thorlabs.com/NewGroupPage9.cfm?ObjectGroup_ID=1001, Jan. 25, 2011.

[97] Vishay Semiconductors, "Datasheet: TEMD5110X01 Silicon PIN Photodiode," May 2009, retrieved from http://www.vishay.com/docs/84658/temd5110.pdf, July 26, 2011.

[98] F.-M. Wu, C.-T. Lin, C.-C. Wei, C.-W. Chen, H.-T. Huang, and C.-H. Ho, "1.1-Gb/s White-LED-Based Visible Light Communication Employing Carrier-Less Amplitude and Phase

Modulation," *IEEE Photonics Technology Letters*, vol. 24, no. 19, pp. 1730–1732, Oct. 2012.

[99] D. Tse and P. Viswanath, *Fundamentals of Wireless Communication*. Cambridge University Press, 2005.

[100] Analog Devices, "Transimpedance Amplifiers," retrieved from http://www.analog.com/en/fiberoptic/transimpedance-amplifiers/products/index.html, Jan. 25, 2011.

[101] B. T. Phong, "Illumination for Computer Generated Pictures," *Communications of the ACM*, vol. 18, no. 6, pp. 311–317, June 1975.

[102] Integra Incorporated, "Specter," retrieved from http://www.integra.jp/en/specter/index.html, Dec. 21, 2008.

[103] T. Whitted, "An Improved Illumination Model for Shaded Display," *Communications of the ACM*, vol. 23, no. 6, pp. 343–349, June 1980.

[104] J. Kajiya, "The Rendering Equation," in *Proceedings of the 13th Annual Conference on Computer Graphics and Interactive Techniques ACM SIGGRAPH 1986*, vol. 20, no. 4, New York, NY, Aug. 1986, pp. 143–150.

[105] S. Pattanaik and S. Mudur, "Computation of Global Illumination by Monte Carlo Simulation of the Particle Model of Light," in *Proceedings of the 3rd Eurographics Workshop on Rendering*, Bristol, May 1992, pp. 71–83.

[106] T. S. Rappaport, *Wireless Communications: Principles and Practice*, 2nd edn. Prentice Hall PTR, 2002.

[107] J. Armstrong, "OFDM for Optical Communications," *Journal of Lightwave Technology*, vol. 27, no. 3, pp. 189–204, Feb. 2009.

[108] H. Elgala, R. Mesleh, and H. Haas, "Practical Considerations for Indoor Wireless Optical System Implementation using OFDM," in *Proceedings of the IEEE 10th International Conference on Telecommunications (ConTel)*, Zagreb, Croatia, Jun. 8–10, 2009.

[109] L. Jingyi, P. Joo, W. Hai, J. Ro, and D. Park, "The Effect of Filling Unique Words to Guard Interval for OFDM System," IEEE 802.16 Broadband Wireless Access Working Group, retrieved May 13, 2009 from http://www.ieee802.org/16/tga/contrib/C80216a-02_87.pdf, Tech. Rep., Sept. 2002.

[110] D. Dardari, V. Tralli, and A. Vaccari, "A Theoretical Characterization of Nonlinear Distortion Effects in OFDM Systems," *IEEE Transactions on Communications*, vol. 48, no. 10, pp. 1755–1764, Oct. 2000.

[111] S. Dimitrov, R. Mesleh, H. Haas, M. Cappitelli, M. Olbert, and E. Bassow, "Line-of-sight Infrared Wireless Path Loss Simulation in an Aircraft Cabin," in *Proceedings of the European Workshop for Photonic Solutions for Wireless, Access, and In-House Networks (IPHOBAC'09)*, Duisburg, Germany, May 18–20, 2009. http://www.ist-iphobac.org/workshop/program.asp.

[112] N. Schmitt, "Wireless optical NLOS Communication in Aircraft Cabin for In-flight Entertainment," in *Proceedings of the ESA 1st Optical Wireless Onboard Communications Workshop*, Noordwijk, Netherlands, Sept. 29–30, 2004.

[113] N. Schmitt, T. Pistner, C. Vassilopoulos, D. Marinos, A. Boucouvalas, M. Nikolitsa, C. Aidinis, and G. Metaxas, "Diffuse Wireless Optical Link for Aircraft Intra-cabin Passenger Communication," in *Proceedings of the 5th International Symposium on Communication Systems, Networks and Digital Signal Processing CSNDSP 2006*, Patras, Greece, July 19–21, 2006.

[114] ATENAA Project, "Advanced Technologies for Networking of Avionic Applications," retrieved from http://www.atenaa.org/, May 8, 2009.

[115] P. Djahani and J. M. Kahn, "Analysis of Infrared Wireless Links Employing Multibeam Transmitters and Imaging Diversity Receivers," *IEEE Transactions on Communications*, vol. 48, no. 12, pp. 2077–2088, Dec. 2000.

[116] F. E. Alsaadi and J. M. H. Elmirghani, "Mobile MC-CDMA Optical Wireless System Employing an Adaptive Multibeam Transmitter and Diversity Receivers in a Real Indoor Environment," in *Proceedings of the IEEE International Conference on Communications (ICC 08)*, May 19–23, 2008, pp. 5196–5203.

[117] D. Wu, Z. Ghassemlooy, H. Le-Minh, S. Rajbhandari, and Y. Kavian, "Power Distribution and Q-factor Analysis of Diffuse Cellular Indoor Visible Light Communication Systems," in the *16th European Conference on Networks and Optical Communications (NOC 2011)*, Newcastle upon Tyne, July 20–22, 2011, pp. 28–31.

[118] D. Wu, Z. Ghassemlooy, H. Le-Minh, S. Rajbhandari, and L. Chao, "Channel Characteristics Analysis of Diffuse Indoor Cellular Optical Wireless Communication Systems," in *Proceedings of the OSA Asia Communications and Photonics Conference and Exhibition (OSA ACP 2011)*, Shanghai, China, Nov. 13–16, 2011, pp. 1–6.

[119] Robert & Associates, "Rhinoceros 3D," retrieved from http://www.rhino3d.com/, Jan. 15, 2009.

[120] Vishay Semiconductors, "Datasheet: TSFF5210 High Speed Infrared Emitting Diode, 870 nm, GaAlAs Double Hetero," retrieved July 26, 2011 from http://www.vishay.com/docs/81090/tsff5210.pdf, June 2009.

[121] —— "Datasheet: TESP5700 Silicon PIN Photodiode, RoHS Compliant," retrieved Dec. 22, 2008 from http://www.vishay.com/docs/81573/tesp5700.pdf, Sept. 2008.

[122] M. Cappitelli, M. Olbert, C. Mussmann, and R. Greule, "GUM in Simulation und Messung," in *Proceedings of Licht2008 Ilmenau*, Ilmenau, Germany, Sept. 10–13, 2008, pp. 487–491.

[123] A. Cheng, S. Rao, L. Cheng, and D. Lam, "Assessment of Pupil Size Under Different Light Intensities Using the Procyon Pupillometer," *Journal of Cataract and Refractive Surgery*, vol. 32, no. 6, pp. 1015–1017, June 2006.

[124] Vishay Semiconductors, "Datasheet: VSMB2020X01 High Speed Infrared Emitting Diodes, RoHS Compliant, 940 nm, GaAlAs, DDH, AEC-Q101 Released," retrieved Dec. 29, 2008 from http://www.vishay.com/docs/81930/vsmb2000.pdf, Nov. 2008.

[125] —— "Datasheet: BPW41N Silicon PIN Photodiode, RoHS Compliant," retrieved Dec. 29, 2008 from http://www.vishay.com/docs/81522/bpw41n.pdf, Sept. 2008.

[126] S. Glisic, *Advanced Wireless Communications: 4G Technologies*. John Wiley & Sons, 2004.

[127] H. Haas and S. McLaughlin, eds., *Next Generation Mobile Access Technologies: Implementing TDD*. Cambridge University Press, Jan. 2008.

[128] P. Omiyi, H. Haas, and G. Auer, "Analysis of TDD Cellular Interference Mitigation Using Busy-Bursts," *Transactions on Wireless Communications*, vol. 6, no. 7, pp. 2721–2731, July 2007.

[129] B. Ghimire, G. Auer, and H. Haas, "Busy Bursts for Trading-off Throughput and Fairness in Cellular OFDMA-TDD," *Eurasip Journal on Wireless Communications and Networking*, vol. 2009, Article ID 462396, 14 pages, 2009.

[130] G. Auer, S. Videv, B. Ghimire, and H. Haas, "Contention Free Inter-Cellular Slot Reservation," *IEEE Communications Letters*, vol. 13, no. 5, pp. 318–320, May 2009.

[131] S. Dimitrov, S. Sinanovic, and H. Haas, "Clipping Noise in OFDM-based Optical Wireless Communication Systems," *IEEE Transactions on Communications (IEEE TCOM)*, vol. 60, no. 4, pp. 1072–1081, April 2012.

[132] —— "A Comparison of OFDM-based Modulation Schemes for OWC with Clipping Distortion," in *GLOBECOM Workshops (GC Wkshps)*, Houston, TX, Dec. 5–9, 2011.

[133] S. Dimitrov and H. Haas, "Information Rate of OFDM-Based Optical Wireless Communication Systems With Nonlinear Distortion," *Journal Lightweight Technology*, vol. 31, no. 6, pp. 918 – 929, March 15, 2013.

[134] S. Dimitrov, "Analysis of OFDM-based Intensity Modulation Techniques for Optical Wireless Communications," Ph.D. dissertation, The University of Edinburgh, Edinburgh, Aug. 2012.

[135] H. Elgala, R. Mesleh, and H. Haas, "Non-linearity Effects and Predistortion in Optical OFDM Wireless Transmission Using LEDs," *Interscience International Journal of Ultra Wideband Communications and Systems (IJUWBCS)*, vol. 1, no. 2, pp. 143–150, 2009.

[136] J. Rice, *Mathematical Statistics and Data Analysis*, 2nd edn. Pacific Grove, CA: Duxbury Press, 1995.

[137] J. Bussgang, "Cross Correlation Function of Amplitude-Distorted Gaussian Signals," Research Laboratory for Electronics, Massachusetts Institute of Technology, Cambridge, MA, Technical Report 216, March 1952.

[138] J. Armstrong and B. J. C. Schmidt, "Comparison of Asymmetrically Clipped Optical OFDM and DC-Biased Optical OFDM in AWGN," *IEEE Communications Letters*, vol. 12, no. 5, pp. 343–345, May 2008.

[139] B. Wilson and Z. Ghassemlooy, "Pulse Time Modulation Techniques for Optical Communications: A Review," *Proceedings of the IEEE on Optoelectronics*, vol. 140, no. 6, pp. 347–357, Dec. 1993.

[140] M. D. Audeh, J. M. Kahn, and J. R. Barry, "Performance of Pulse-position Modulation on Measured Non-directed Indoor Infrared Channels," *IEEE Transactions on Communications*, vol. 44, no. 6, pp. 654–659, June 1996.

[141] D. Lee, J. Kahn, and M. Audeh, "Trellis-Coded Pulse-Position Modulation for Indoor Wireless Infrared Communications," *IEEE Transactions on Communications*, vol. 45, no. 9, pp. 1080–1087, Sept. 1997.

[142] S. Randel, F. Breyer, S. C. J. Lee, and J. W. Walewski, "Advanced Modulation Schemes for Short-Range Optical Communications," *IEEE Journal of Selected Topics in Quantum Electronics*, vol. 16, no. 5, pp. 1280–1289, 2010.

[143] Y. Zeng, R. Green, and M. Leeson, "Multiple Pulse Amplitude and Position Modulation for the Optical Wireless Channel," in *Proceedings of the 10th Anniversary International Conference on Transparent Optical Networks (ICTON'08)*, vol. 4, Athens, Greece, June 22–26, 2008, pp. 193–196.

[144] J. Barry and J. Kahn, "Design of Non-Directed Infrared Links for High-Speed Wireless Networks," in *Proceedings of the Lasers and Electro-Optics Society Annual Meeting LEOS 1994*, Boston, MA, Oct. 31–Nov 3. 1994.

[145] M. Akbulut, C. Chen, M. Hargis, A. Weiner, M. Melloch, and J. Woodall, "Digital Communications Above 1 Gb/s Using 890-nm Surface-Emitting Light-Emitting Diodes," *IEEE Photonics Technology Letters*, vol. 13, no. 1, pp. 85–87, Jan. 2001.

[146] S. Rajbhandari, Z. Ghassemlooy, and M. Angelova, "Optimising the Performance of Digital Pulse Interval Modulation with Guard Slots for Diffuse Indoor Optical Wireless Links," *IET Microwaves, Antennas and Propagation*, vol. 5, no. 9, pp. 1025–1030, June 2011.

[147] J. B. Carruthers and J. M. Kahn, "Multiple-subcarrier Modulation for Nondirected Wireless Infrared Communication," *IEEE Journal on Selected Areas in Communications*, vol. 14, no. 3, pp. 538–546, April 1996.

[148] T. Ohtsuki, "Multiple–Subcarrier Modulation in Optical Wireless Communications," *IEEE Communications Magazine*, vol. 41, no. 3, pp. 74–79, March 2003.

[149] P. Golden, H. Dedieu, and K. Jacobsen, *Fundamentals of DSL Technology*. Auerbach Publications, 2006.

[150] B. Inan, S. C. J. Lee, S. Randel, I. Neokosmidis, A. M. J. Koonen, and J. W. Walewski, "Impact of LED Nonlinearity on Discrete Multitone Modulation," *IEEE/OSA Journal of Optical Communications and Networking*, vol. 1, no. 5, pp. 439 –451, Oct. 2009.

[151] H. Elgala, R. Mesleh, and H. Haas, "Impact of LED Nonlinearities on Optical Wireless OFDM Systems," in *2010 IEEE 21st International Symposium on Personal Indoor and Mobile Radio Communications (PIMRC)*, Sept. 2010, pp. 634 – 638.

[152] K. Asadzadeh, A. Dabbo, and S. Hranilovic, "Receiver Design for Asymmetrically Clipped Optical OFDM," in *GLOBECOM Workshops (Optical Wireless Communication), 2011 IEEE*, Houston, TX, Dec. 5–9, 2011, pp. 777–781.

[153] Q. Pan and R. J. Green, "Bit-Error-Rate Performance of Lightwave Hybrid AM/OFDM Systems with Comparison with AM/QAM Systems in the Presence of Clipping Impulse Noise," *IEEE Photonics Technology Letters*, vol. 8, no. 2, pp. 278–280, Feb. 1996.

[154] H. Ochiai and H. Imai, "Performance Analysis of Deliberately Clipped OFDM Signals," *IEEE Transactions on Communications*, vol. 50, no. 1, pp. 89–101, Jan. 2002.

[155] A. Bahai, M. Singh, A. Goldsmith, and B. Saltzberg, "A New Approach for Evaluating Clipping Distortion in Multicarrier Systems," *IEEE Journal on Selected Areas in Communications*, vol. 20, no. 5, pp. 1037–1046, June 2002.

[156] D. J. G. Mestdagh, P. Spruyt, and B. Biran, "Analysis of Clipping Effect in DMT-based ADSL Systems," in *Proceedings of the IEEE International Conference on Communications ICC 1994*, vol. 1, New Orleans, LA, May 1–5, 1994, pp. 293–300.

[157] E. Vanin, "Signal Restoration in Intensity-Modulated Optical OFDM Access Systems," *OSA Optics Letters*, vol. 36, no. 22, pp. 4338–4340, 2011.

[158] S. C. J. Lee, F. Breyer, S. Randel, H. P. A. van der Boom, and A. M. J. Koonen, "High-speed Transmission over Multimode Fiber Using Discrete Multitone Modulation," *Journal of Optical Networking*, vol. 7, no. 2, pp. 183–196, Feb. 2008.

[159] R. J. Green, H. Joshi, M. D. Higgins, and M. S. Leeson, "Recent Developments in Indoor Optical Wireless," *IET Communications*, vol. 2, no. 1, pp. 3–10, Jan. 2008.

[160] J. Li, X. Zhang, Q. Gao, Y. Luo, and D. Gu, "Exact BEP Analysis for Coherent M-arry PAM and QAM over AWGN and Rayleigh Fading Channels," in *Proceedings of the IEEE Vehicular Technology Conference (VTC 2008-Spring)*, Singapore, May 11–14, 2008, pp. 390–394.

[161] N. Johnson, S. Kotz, and N. Balakrishnan, *Continuous Univariate Distributions*, 2nd edn. John Wiley & Sons Ltd., 1994, vol. 1.

[162] P. K. Vitthaladevuni, M.-S. Alouini, and J. C. Kieffer, "Exact BER Computation for Cross QAM Constellations," *IEEE Transactions on Wireless Communications*, vol. 4, no. 6, pp. 3039–3050, Nov. 2005.

[163] J. Smith, "Odd-Bit Quadrature Amplitude-Shift Keying," *IEEE Transactions on Communications*, vol. 23, no. 3, pp. 385–389, March 1975.

[164] S. Dimitrov and H. Haas, "On the Clipping Noise in an ACO-OFDM Optical Wireless Communication System," in *Proceedings of the IEEE Global Communications Conference (IEEE GLOBECOM 2010)*, Miami, FL, Dec. 6–10, 2010.

[165] C. Shannon, "A Mathematical Theory of Communication," *Bell System Technical Journal*, vol. 27, pp. 379–423 and 623–656, July and Oct. 1948.

[166] G. D. Forney and G. Ungerboeck, "Modulation and Coding for Linear Gaussian Channels," *IEEE Transactions on Information Theory*, vol. 44, no. 6, pp. 2384–2415, Oct. 1998.

[167] R.-J. Essiambre, G. Kramer, P. Winzer, G. Foschini, and B. Goebel, "Capacity Limits of Optical Fibre Networks," *IEEE/OSA Journal on Lightwave Technology (IEEE/OSA JLT)*, vol. 28, no. 4, pp. 662–701, Feb. 15, 2010.

[168] S. Hranilovic and F. Kschischang, "Capacity Bounds for Power- and Band-limited Optical Intensity Channels Corrupted by Gaussian Noise," *IEEE Transactions on Information Theory*, vol. 50, no. 5, pp. 784–795, May 2004.

[169] C. Shannon, "Communication in the Presence of Noise," in *Proceedings of the IRE*, vol. 37, no. 1, pp. 10–21, Jan. 1949.

[170] A. Farid and S. Hranilovic, "Capacity of Optical Intensity Channels with Peak and Average Power Constraints," in *Proceedings of the IEEE International Conference on Communications (IEEE ICC 2009)*, Dresden, Germany, June 14–18, 2009, pp. 1–5.

[171] —— "Capacity Bounds for Wireless Optical Intensity Channels with Gaussian Noise," *IEEE Transactions on Information Theory*, vol. 56, no. 12, pp. 6066–6077, Dec. 2010.

[172] R. You and J. Kahn, "Upper-bounding the Capacity of Optical IM/DD Channels with Multiple-subcarrier Modulation and Fixed Bias Using Trigonometric Moment Space Method," *IEEE Transactions on Information Theory*, vol. 48, no. 2, pp. 514–523, Feb. 2002.

[173] X. Li, R. Mardling, and J. Armstrong, "Channel Capacity of IM/DD Optical Communication Systems and of ACO-OFDM," in *Proceedings of the IEEE International Conference on Communications (IEEE ICC 2007)*, Glasgow, June 24–28 2007, pp. 2128–2133.

[174] X. Li, J. Vucic, V. Jungnickel, and J. Armstrong, "On the Capacity of Intensity-Modulated Direct-Detection Systems and the Information Rate of ACO-OFDM for Indoor Optical Wireless Applications," *IEEE Transactions on Communications (IEEE TCOM)*, vol. 60, no. 3, pp. 799–809, March 2012.

[175] J. Tellado, L. M. C. Hoo, and J. M. Cioffi, "Maximum-Likelihood Detection of Nonlinearly Distorted Multicarrier Symbols by Iterative Decoding," *IEEE Transactions on Communications*, vol. 51, no. 2, pp. 218–228, Feb. 2003.

[176] I. Gutman and D. Wulich, "On Achievable Rate of Multicarrier with Practical High Power Amplifier," in *Proceedings of the European Wireless Conference (EW 2012)*, Poznan, Poland, April 18–20, 2012, pp. 1–5.

[177] D. Kim and G. L. Stueber, "Clipping Noise Mitigation for OFDM by Decision-Aided Reconstruction," *IEEE Communications Letters*, vol. 3, no. 1, pp. 4–6, Jan. 1999.

[178] H. Chen and A. M. Haimovich, "Iterative Estimation and Cancellation of Clipping Noise for OFDM Signals," *IEEE Communications Letters*, vol. 7, no. 7, pp. 305–307, July 2003.

[179] S. Boyd and L. Vandenberghe, *Convex Optimization*. Cambridge University Press, 2004.

[180] MathWorks, "Matlab Documentation Center," retrieved from http://www.mathworks.co.uk/help/matlab/index.html, Nov. 17, 2012.

[181] M. Dyble, N. Narendran, A. Bierman, and T. Klein, "Impact of Dimming White LEDs: Chromaticity Shifts Due to Different Dimming Method," in *Proceedings of the SPIE 5941, 5th International Conference on Solid State Lighting*, Bellingham, WA, 2005, pp. 291–299.

[182] Y. Gu, N. Narendran, T. Dong, and H. Wu, "Spectral and Luminous Efficacy Change of High-Power LEDs under Different Dimming Methods," in *Proceedings of the SPIE 6337, 6th International Conference on Solid State Lighting*, San Diego, CA, 2006.

[183] K. Loo, Y. Lai, S. Tan, and C. Tse, "On the Color Stability of Phosphor-Converted White LEDs under DC, PWM, and Bi-Level Drive," *IEEE Transactions on Power Electronics*, vol. 2, pp. 974–984, 2012.

[184] E. Telatar, "Capacity of Multi-Antenna Gaussian Channels," *European Transactions on Telecommun.*, vol. 10, no. 6, pp. 585–595, Nov./Dec. 1999.

[185] G. J. Foschini and M. J. Gans, "On Limits of Wireless Communications in a Fading Environment when Using Multiple Antennas," *Wireless Personal Communications*, vol. 6, no. 6, pp. 311–335, 1998.

[186] T. Fath, "Evaluation of Spectrally Efficient Indoor Optical Wireless Transmission Techniques," Ph.D. Thesis, The University of Edinburgh, 2013, http://www.eng.ed.ac.uk/drupal/hxh/downloads/.

[187] S. G. Wilson, M. Brandt-Pearce, Q. Cao, and M. Baedke, "Optical Repetition MIMO Transmission with Multipulse PPM," *IEEE Journal on Selected Areas in Communications*, vol. 23, no. 9, pp. 1901–1910, Sept. 2005.

[188] S. M. Navidpour, M. Uysal, and M. Kavehrad, "BER Performance of Free-Space Optical Transmission with Spatial Diversity," *IEEE Transactions on Wireless Communications*, vol. 6, no. 8, pp. 2813–2819, Aug. 2007.

[189] D. O'Brien, "Multi-Input–Multi-Output (MIMO) Indoor Optical Wireless Communications," in *Conference Record of the Forty-Third Asilomar Conference on Signals, Systems and Computers*, Nov. 2009, pp. 1636–1639.

[190] L. Zeng, D. O'Brien, H. Minh, G. Faulkner, K. Lee, D. Jung, Y. Oh, and E. T. Won, "High Data Rate Multiple Input Multiple Output (MIMO) Optical Wireless Communications using White LED Lighting," *IEEE Journal on Selected Areas in Communications*, vol. 27, no. 9, pp. 1654–1662, Dec. 2009.

[191] T. Fath and H. Haas, "Performance Comparison of MIMO Techniques for Optical Wireless Communications in Indoor Environments," *IEEE Transactions on Communications*, vol. 61, pp. 733–742, 2013.

[192] Y. A. Chau and S.-H. Yu, "Space Modulation on Wireless Fading Channels," in *Proceedings of the IEEE Vehicular Technology Conference (VTC Fall 2001)*, vol. 3, Oct. 7–11, 2001, pp. 1668–1671.

[193] R. Mesleh, H. Haas, S. Sinanović, C. W. Ahn, and S. Yun, "Spatial Modulation," *IEEE Transactions on Vehicular Technology*, vol. 57, no. 4, pp. 2228–2241, July 2008.

[194] R. Mesleh, H. Elgala, and H. Haas, "Optical Spatial Modulation," *IEEE/OSA Journal of Optical Communications and Networking*, vol. 3, no. 3, pp. 234–244, March 2011.

[195] M. Di Renzo, H. Haas, A. Ghrayeb, S. Sugiura, and L. Hanzo, "Spatial Modulation for Generalized MIMO: Challenges, Opportunities, and Implementation," *Proceedings of the IEEE*, vol. 102, no. 1, pp. 56–103, Jan. 2014.

[196] T. Fath and H. Haas, "Optical Spatial Modulation Using Colour LEDs," in *Proceedings of the IEEE International Conference on Communications (ICC)*, June 9–13, 2013, pp. 3938–3942.

[197] O. Bouchet, G. Faulkner, L. Grobe, E. Gueutier, K. D. Langer, S. Nerreter, D. O'Brien, R. Turnbull, J. Vucic, J. Walewski, and M. Wolf, "Deliverable D4.2b Physical Layer Design and Specification, Seventh Framework Programme Information and Communication Technologies," retrieved July 18, 2012 from http://www.ict-omega.eu, Feb. 2011.

[198] D. O'Brien, G. Faulkner, H. Le Minh, O. Bouchet, M. El Tabach, M. Wolf, J. Walewski, S. Randel, S. Nerreter, M. Franke, K. D. Langer, J. Grubor, and T. Kamalakis, "Home

Access Networks Using Optical Wireless Transmission," in *Proceedings of the IEEE International Symposium on Personal, Indoor and Mobile Radio Communications*, Cannes, France, Sept. 15–18, 2008, pp. 1–5.

[199] M. Safari and M. Uysal, "Do We Really Need OSTBCs for Free-Space Optical Communication with Direct Detection?" *IEEE Transactions on Wireless Communications*, vol. 7, no. 11, pp. 4445–4448, Nov. 2008.

[200] S. M. Alamouti, "A Simple Transmit Diversity Technique for Wireless Communications," *IEEE Journal on Selected Areas in Communications*, vol. 16, no. 8, pp. 1451–1458, Oct. 1998.

[201] S. Hranilovic, *Wireless Optical Communication Systems*, 1st edn. Springer, Sept. 1996.

[202] A. Goldsmith, *Wireless Communications*. Cambridge University Press, 2005.

[203] R. Mesleh, M. Di Renzo, H. Haas, and P. M. Grant, "Trellis Coded Spatial Modulation," *IEEE Transactions on Wireless Communications*, vol. 9, no. 7, pp. 2349–2361, July 2010.

[204] E. Basar, U. Aygolu, E. Panayirci, and V. H. Poor, "Space-Time Block Coded Spatial Modulation," *IEEE Transactions on Communications*, vol. 59, no. 3, pp. 823–832, March 2011.

[205] R. Mesleh, H. Haas, C. W. Ahn, and S. Yun, "Spatial Modulation: A New Low Complexity Spectral Efficiency Enhancing Technique," in *Proceedings of the IEEE International Conference on Communication and Networking in China (CHINACOM)*, Beijing, China, Oct. 25–27, 2006, pp. 1–5.

[206] R. Mesleh, "Spatial Modulation: A Spatial Multiplexing Technique for Efficient Wireless Data Transmission," Ph.D. dissertation, Jacobs University, Bremen, Germany, June 2007.

[207] A. Younis, M. Di Renzo, R. Mesleh, and H. Haas, "Sphere Decoding for Spatial Modulation," in *Proceedings of the IEEE International Conference on Communications (ICC)*, Kyoto, Japan, June 5–9, 2011, pp. 1–6.

[208] K.-D. Langer, J. Vucic, C. Kottke, L. Fernandez, K. Habel, A. Paraskevopoulos, M. Wendl, and V. Markov, "Exploring the Potentials of Optical-Wireless Communication Using White LEDs," in *Proceedings of the International Conference on Transparent Optical Networks (ICTON)*, June 26–30, 2011.

[209] H. Uchiyama, M. Yoshino, H. Saito, M. Nakagawa, S. Haruyama, T. Kakehashi, and N. Nagamoto, "Photogrammetric System Using Visible Light Communications," in *Proceedings of the Annual Conference of IEEE Industrial Electronics (IECON)*, Orlando, FL: IEEE, Nov. 10–13, 2008, pp. 1771–1776.

[210] T. Y. H. Okada, K. Masuda, and M. Katayama, "Successive Interference Cancellation for Hierarchical Parallel Optical Wireless Communication Systems," in *Proceedings of the Asia-Pacific Conference on Communications*, Perth, USA: IEEE, Oct. 25–28, 2005, pp. 788–792.

[211] 3GPP, "Evolved Universal Terrestrial Radio Access (E-UTRA); Radio Frequency (RF) System Scenarios," 3GPP TR 36.942 V 8.2.0 (2009-05), retrieved Dec. 1, 2009 from www.3gpp.org/ftp/Specs/, May 2009.

[212] Y. Argyropoulos, S. Jordan, and S. P. R. Kumar, "Dynamic Channel Allocation in Interference-Limited Cellular Systems with Uneven Traffic Distribution," *IEEE Transactions on Vehicular Technology*, vol. 48, no. 1, pp. 224–232, Jan. 1999.

[213] K. I. X. Lin and K. Hirohashi, "High-Speed Full-Duplex Multiaccess System for LEDs Based Wireless Communications Using Visible Light," in *Proceedings of the International*

Symposium on Optical Engineering and Photonic Technology (OEPT), Orlando, FL, July 10–13, 2009.

[214] Osram Opto Semiconductors, "SFH 4730, Lead (Pb) Free Product–RoHS Compliant," retrieved from http://catalog.osram-os.com, March 25, 2011.

[215] Hamamatsu Photonics K.K., "Datasheet: Si PIN photodiode S6801/S6898 series," retrieved Oct. 18, 2010 from http://sales.hamamatsu.com/assets/pdf/parts_S/S6801_etc.pdf, Sept. 2008.

[216] B. Ghimire, G. Auer, and H. Haas, "Heuristic Thresholds for Busy Burst Signalling in a Decentralised Coordinated Multipoint Network," in *Proceedings of the 73rd IEEE Vehicular Technology Conference (VTC)*, Budapest, Hungary: IEEE, May 15–19, 2011, pp. 5 pages on CD–ROM.

Index

Printed in the United States
By Bookmasters